EUREKA!
The Book of Inventing

EUREKA!

The Book of Inventing

Bob Symes
and
Robin Bootle

HEADLINE

First published in 1994
by HEADLINE BOOK PUBLISHING

10 9 8 7 6 5 4 3 2 1

British Library Cataloguing in Publication Data

Symes, Bob
Eureka! Book of Inventing
I. Title II. Bootle, Robin
600

ISBN 0-7472-1147-7

Typeset by
Letterpart Limited, Reigate, Surrey

Printed and bound in Great Britain by
Mackays of Chatham PLC, Chatham, Kent

HEADLINE BOOK PUBLISHING
A division of Hodder Headline PLC
338 Euston Road
London NW1 3BH

To the many inventors who nearly got there.

And to Valerie.

Contents

Acknowledgements

The authors would like to express their gratitude to all those who helped in the writing of this book.

Especial thanks to Alan Wilcher of Imagineering, patent attorney Graham H. Jones, the Institute of Patentees and Inventors for helping to organize a survey of its members, and the members themselves, who took so much trouble to let us know about their triumphs and failures, their hopes and their fears.

We have made every effort to ensure that the book's examples and advice are accurate. This is not a legal textbook and, as suggested in the text, you should confirm points of law with your own advisers.

Bob Symes'
Ten Rules for Inventors

Rule One
Identify the problem

Rule Two
Meet a need

Rule Three
Keep on learning

Rule Four
Check for originality

Rule Five
Build a working model

Rule Six
Don't attack established interests

Rule Seven
Learn the patent system

Rule Eight
Find a product champion

Rule Nine
Sell yourself as well as the invention

Rule Ten
Persevere

CHAPTER ONE

Can *You* Be an Inventor?

Yes, you can. Almost anyone can. Your talent may be hidden, but I am going to show you how to use it.

Push aside any blockage in your mind which says that inventors are either mad or geniuses or both. They aren't. Inventing is a profession like any other. You need an amount of flair to get ahead, of course, but real success comes from hard work and practice.

Among hundreds of new ideas, Thomas Alva Edison came up with the phonograph (ancestor of today's compact disc player) and the electric lamp. He worked – how he worked – and he said it better than I can: invention is one per cent inspiration and ninety-nine per cent perspiration. Forget the perspiration for the moment – we'll come back to that. Your first aim is to understand that you contain within yourself that magic one per cent of inspiration.

The British have a talent for invention. Japan's Ministry of International Trade and Industry states that more than half the successful inventions since the last war began life in the UK. Only a fifth came from the USA and a paltry twentieth from Japan itself.

Where the UK goes wrong is in the way it fails to exploit new ideas, leaving them to other countries to develop. The usual reaction to this problem is much public wailing and gnashing of teeth from 'experts'. They ignore the fact that many of the inventors involved actually *made* money – by licensing or selling their products abroad.

It's almost impossible to adjust the national psyche – to change a nation of inventors into thrusting businessmen. But why waste time on such a fruitless exercise when we can more easily improve an area where we are already world leaders?

Inventing. That's why I have written this book.

My aim is to inspire you and give you clues to successful innovation. I start with how to dream up ideas and translate them into reality, move on through problems of patenting and approaching manufacturers, and sign off as you sell your invention and cash registers ring up their tribute to success.

My own progress as a slightly successful inventor illustrates many of my arguments. If only I had known, as I lived through events, that I was providing my own case study. If only someone had written this book for me thirty years ago! Some things might have turned out differently.

But I always try to learn. As President of the Institute of Patentees and Inventors (having been formerly its Chairman) I am continually involved with our members, who are some of the most active innovators in the country.

I spent fifteen years as Unit Co-ordinator of BBC television's *Tomorrow's World*, which meant, among other things, that I handled thousands of viewers' questions about invention. They taught me a lot.

Since then I've had a further fifteen years of the pleasure of working with innovators, presenting their ideas on television, judging their work at engineering exhibitions and innovation contests and, behind the scenes, trying to help them with suggestions for improvements.

I have met many inventors, and I can honestly say that hardly any of them complain of difficulty in thinking up new products or new ideas. You, too, can be an inventor! You don't have to be brilliant. I don't have a 'superior' education. I didn't go to university, and I never worked for a giant industrial company with access to research laboratories and equipment costing hundreds of thousands of pounds. I am a private citizen. But I innovate – it's second nature to me now.

We all invent without realizing it. An everyday activity like cooking becomes invention the moment you break away from the recipe book. Set the cooker at a higher temperature than usual to produce a piquant caramel effect, and you are inventing. Try a new mixture of herbs to change a dish's flavour. Again you have an invention. Or rather you have an innovation. The word 'invention' applies to objects which can be made; an innovation is a new idea.

It is impossible to tell in advance where innovation turns into invention. But what is clear is that sticking to the second word narrows your view hopelessly because, when your mind should be happily freewheeling through fresh thoughts, it snags on premature worries about patents and raising money. So I prefer to talk of innovators rather than inventors.

Let's start your career as an innovator now. Do you realize that you have probably already found your own original ways of solving problems in your home? You may have done it when you planned a fresh layout for a room, or perhaps devised a game to play with your children, or designed your own version of a compost heap. It's not difficult.

The innovating spark exists in all of us. It is no great step, when you find that you can't buy a product to solve a problem, to design the answer yourself. Which brings us to:

Rule One: Identify the problem

Although I could go to engineering or medicine or agriculture for examples of 'identifying the problem', you may not be familiar with such areas, which can call for specialist knowledge. But if we think about the average house or flat, we should be on common ground. Our homes are full of possibilities for improvement, provided that you recognize them.

It is a mistake to assume the devices you use in the kitchen or garage have been around for so long that nobody can improve them any more. It's not so. Start with your own daily experience and ask: 'Do they have to do it that way?' Your motto must be that *anything* can be made to work better. And it can. Let me prove it to you with three simple examples.

When you think of the untold millions of farm labourers and gardeners who have wielded spades over the course of history, your heart may sink. Surely all their experience and skill must have produced the perfect digging implement? How can something as elementary as a spade be made better? It doesn't seem possible, does it? The traditional spade is a superb tool with the advantage of being extremely simple, both in its design and use.

But if you aren't skilled in handling a spade, a heavy digging session can give you a painfully damaged back. And if you are already weakened – bad backs permanently affect at least one in ten of us – you cannot dig at all. The spring-loaded spade was successful because its inventor *identified this problem* before solving it.

How does it work? You thrust the heavy spade straight down into the ground, partly under its own weight, which saves you effort to begin with. As you pull the spade handle back – using arm-work, not a bent back – the energy exerted is stored in the spring. That energy is then automatically released to lift the spit of earth clear of the ground and throw it forward.

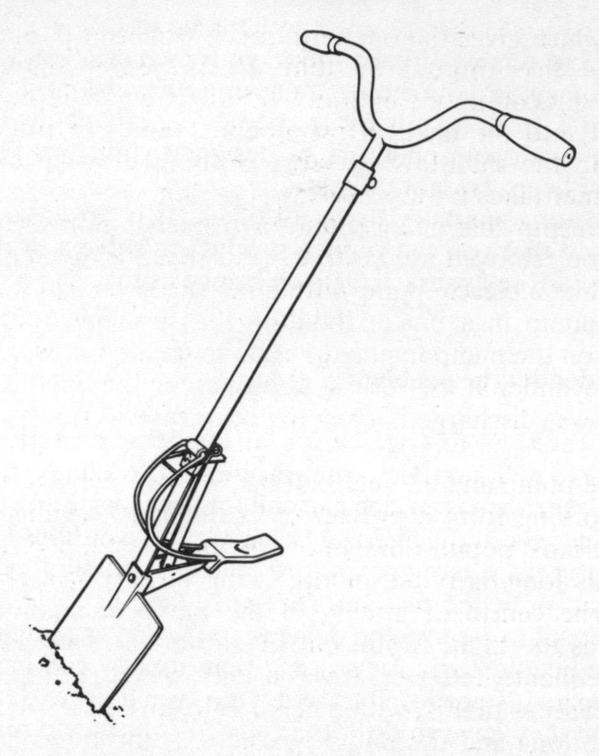

What a brilliant idea it is! Your back stays straight throughout – making you into a more efficient machine yourself – and the effort once spent on digging is now reused to lift the soil, halving the workload. People who could not use a conventional spade are able to garden again.

My second example is the tin opener. Do you remember the standard British version, a fearsome object with a sharp metal blade ending in a spike? You dug the pointed part into the top of the tin, and worked the blade round with a sawing motion until a disc of metal came loose. Few users failed at one time or other to bloody their fingers on the jagged edges which the opener created round the rim of the tin.

The design was first improved by the invention of a device with a cutting wheel at the centre of two levered arms. Squeezing the arms forced the wheel into the metal of the top of the tin. Turning a key next to the cutter then drove the opener around its top until the tin was opened. But some problems were not solved: it was hard work to cut the thick top from the tin, and the wheel became blunt with use.

Today's tin opener is successful because its inventor *identified this problem* before solving it. It works very simply, by turning the earlier tool on its side. The levers now lock on to the tin's raised rim, which gives the opener a much firmer grip, and the cutting wheel slices through the thin *side* of the tin. The entire top of the tin lifts off in one motion, with the aid of a permanent magnet built into the head of the opener. You can run your finger around the smoothly cut edge without drawing blood. And the opener takes far less effort to operate.

The garden spray has changed out of all recognition since the days when you removed aphids from your roses using an object that looked like a bicycle pump with a nozzle on the front. You dipped the pump in a bucket holding the spraying fluid and hauled back on the pump handle to suck up the liquid. Next you levelled the cylinder at the roses and pushed on the handle until all the spray was discharged. Then the business was repeated all over again.

Joining the pump and an enclosed bucket together – to make a sprayer working from a pressurized container – changed all that. But the most popular design on the market still had faults. Worst was its long-barrelled pump, rising on top of a conical container. The centre of gravity of the pump and container combined was too high. If you put the spray down on uneven ground, it frequently fell over. It was also difficult to pump up to pressure 'cleanly' – that is, with nice straight strokes – when the barrel was so long and the pump handle rose three feet above the ground at maximum extension.

The redesigner of the spray was successful because he *identified this problem* before solving it. His answer was so simple, so elegant, that it almost makes me weep for pleasure to look at it. He put the pump barrel inside the container instead of on top of it! This makes the spray easier to pump and much less likely to fall over, because its weight is closer to the ground. Add a couple more refinements, such as making the container cylindrical – this helps stability by increasing the amount of liquid the container can hold – and a pressure valve, and you have a completely redesigned garden spray.

These are not fusty examples of Victorian invention. They are well-known, tried-and-tested, household objects which have nonetheless been massively improved in the late twentieth century. The crucial thing to remember is that none of them came about by accident. They all needed human beings to stop and ask the question: 'Can it be done better?'

Why shouldn't the next person to create something which will improve our lives be you? Look around for difficulties that need solving. Identify them. They are there to be found, once you start looking.

Here are a few problems and answers from the last ten years.

The solutions aren't all patentable and some of them won't make the innovators a fortune, but they do show that the potential for invention, in the domestic field alone, is endless.

- *Problem: radiators drip messily when they are bled to release airlocks in central-heating systems.*
Answer: a bottle with a key in its neck which opens the valve and bleeds water into the bottle, instead of letting it drip on the floor or the carpet.

- *Problem: banging cat flaps.*
Answer: a flap mounted on a *sideways* central pivot. The aerodynamics of its design prevent it blowing open and shut in a breeze, while the cat can still go freely in and out.

- *Problem: curtains are hard to handle when you want to wash them or redecorate.*
Answer: a mounting which lets curtains and rail be lowered as one unit for easy handling and rehanging.

- *Problem: how can blind or partially sighted people know when their bath is full?*
Answer: an alarm which measures the depth of water and gives an audible warning when there is enough in the bath.

I don't want to give the impression that I am interested only in lightweight domestic ideas, even though there is money to be made from many of them. Personally, I prefer rather bigger game. But I've concentrated on household innovations to make a point.

None of the problems I have listed in this chapter needed the slightest amount of technical knowledge to recognize, and only the cat flap and the spring-loaded spade required any know-how to contrive a solution.

Almost anyone could have come up with ideas like these. The people who devised them – they include a schoolgirl – don't have two heads. They are normal everyday folk. With this book's advice, you can hope to join or surpass them. With determination and practice, it can come true. You *can* be an inventor.

CHAPTER TWO

Teach Yourself to Invent

There is only one way to learn to invent – and that's to do it. 'Ridiculous!' do I hear you cry? If you could invent just like that, you wouldn't be reading this book. But I don't mean you to build an invention – not just yet, anyway. I want you to practise innovation in your head, as a form of training.

One of my pleasures is constructing engineering models, especially railway locomotives. I don't think about them only when I switch on the lathe in my workshop; at a hidden level, they occupy part of my mind all the time. I cannot visit a railway anywhere – full-scale or model – without taking mental notes. Long ago, I trained myself to think this way. Question after question passes through my mind. How does that injector work? Could that coupling be improved?

The pay-off for this highly enjoyable exercise is that occasionally the lightning of creation strikes. If I hadn't trained for many years, I would never have beaten the problems of designing the world's first working diesel-electric model locomotive or of constructing a model railway carriage with a unique powered bogey.

You too must practise, until you find it difficult to stop wondering, 'Can I improve it?'; until innovation never leaves your mind. You might be lucky and come up with a winner the first time you try, which brings us to an important point.

Write It Down

It's amazing how easy it is to forget a bright thought, especially when it's only part of an overall solution you are looking for. Any interruption can make you lose your thread. And once you have lost it, because by definition your thought was new there is nowhere you can go to look it up.

Take warning from what happened to the poet Coleridge.

Asleep at his Somerset home, he dreamed the whole of his famous poem 'Kubla Khan'. On waking, he began feverishly writing it out but was interrupted by a visitor. The unwanted guest stayed with him until he no longer recalled the second half of the poem. It remains unfinished, with only the brilliant imagery of its existing verses as evidence of what he, and we, lost.

There are no words to describe the frustration of forgetting. The answer is to keep a little daybook in which you set down anything connected with innovating. Don't worry if inspiration strikes at an inconvenient time – just jot a few key words on the edge of a newspaper or a cigarette packet or a Post-it (now there's a simple but shatteringly successful idea). Suppose you've been thinking that a toothbrush is unhygienic. Why not have a brush you can throw away after use? Two words – 'disposable toothbrush' – will remove the danger of your memory lapsing.

Let's get down to the training programme. I'm going to lead you through a simple exercise in thinking, using principles you can apply over and over again. Basically it consists of asking yourself questions in a fairly logical order. Most of them lead to blind corners and halts but you must expect that.

Take 'improving the vacuum cleaner' as an example of the way your thought track might work. Vacuum cleaners have undergone nearly 100 years of development. It may seem impossible to make them better. But is this really so?

Remember Rule One: *Identify the problem*.

Ask yourself what don't you like about vacuum cleaners? One complaint leaps to mind. They are usually underpowered, so that it takes several passes to clean one section of floor. Existing cleaners rely on no more than a large fan. This sucks the air – at low vacuum – through the bag which traps the dust. As dust builds up in the bag, the efficiency of the fan drops off.

One way to improve the situation would be a larger motor to drive the fan. But, you think to yourself, that would put up the cleaner's price, making it difficult to sell unless you find a specialized market.

Your research reveals that this has been done. There is already at least one high-power cleaner available and it does sell to a special-interest group. It is advertised to help people with asthma because it retains more house dust than usual and also sucks up tiny allergy-inducing lifeforms.

The competition is in possession of the field. You've run into your first dead end. What other ideas can you try? How about making the vacuum without a fan at all? Now maybe you are getting somewhere. Why not use a vortex chamber, which is a type of centrifuge, a spinning cylinder?

Could it work? It might. The method is well known; timber mills and carpenters call it the 'cyclone' and use it to remove sawdust and shavings from the workshop floor.

Look at a sawmill and you will find a large metal cone high on the wall. This, the vortex chamber, drops dust and shavings into a collector at ground level while expelling waste air through its top.

Using the same principle in a vacuum cleaner, could dirt be pushed downwards while air escapes the other way? No bag would lie between the source of vacuum and the carpet which it is cleaning. So the machine would be much more efficient and stay at that efficiency. This idea might be worth pursuing.

Write it down: 'Vacuum – vortex.'

But you can't stop there. You must consider the add-on effects of using higher vacuum. You will need some sort of spoiler for the brush heads, to stop their extra suction lifting carpets from the floor or eating cushion covers and soft furnishings. Hoses must be strong so that they do not collapse under the increased suction. But then they will not be so flexible, and you will have difficulty using them. How can this problem be solved?

Another problem with conventional cleaners occurs to you. When they are towed around at the end of the hose, they bang into furniture and chip the paintwork. What can be done about this?

You see that by starting with one basic question – is the vacuum cleaner as good as it might be? – you are drawn on through a whole series of further thoughts which might eventually lead to the design of an improved machine. If you practise this technique, you will find that it becomes easier to think of fresh ideas. You are preparing yourself. Stories of 'chance' discoveries are legion. Many of them were not chance at all, but simply revealed themselves to those with prepared and inquiring minds.

Ron Hickman found the mental track which led him to invent his famous Workmate folding workbench when he accidentally sawed through a chair he was using to support some carpentry. But even then I doubt that he would have come up with the Workmate without years of problem-solving experience in engineering design behind him.

The approach I am urging on you can be used over and over again. For a second trial run, let's look at the domestic refrigerator. The heat which a fridge takes from inside itself to keep its contents cold is extracted by a heat exchanger. It warms the room where the fridge stands – the kitchen, utility room or garage. The problem is that in the first two locations, the heat is not needed; in the garage it is certainly wasted. In our energy-conscious age, can that heat be saved or used in some way?

For instance, could you link a fridge by small-bore piping to a separate heat exchanger placed in the airing cupboard? You would be using the waste energy in a productive way to air clothes. But that doesn't work. Most airing cupboards don't need warming because they already hold a hot-water cylinder. There is no problem – and therefore no need for a solution. Innovation involves many blind alleys and this is one of them.

Don't give up. Think about how else the idea might be used. By keeping a room in the house warm? No. There isn't enough heat available from the fridge. Then how about heating a greenhouse? This might ease the gardener's perennial problem of coddling his tender plants through the winter. Put the heat exchanger in the greenhouse.

Well done. You've identified the problem – fridges waste heat – and you've found a way to use that heat. But will anyone pay good money for your invention?

Rule Two: Meet a need

Never forget that people buy for their own satisfaction, not to improve the state of inventors' bank balances or to compliment them on how clever they are. Consumers please themselves first. You know they do, because you are a consumer yourself. The best way to persuade buyers to dig out their charge cards is to give them what they want. Meet their demand, even if they themselves do not yet know what it is. That is where a profitable future lies: in an imaginative understanding of what people require.

Charge cards are themselves a classic example of how to satisfy consumers' unspoken needs. Businessman Frank McNamara was entertaining a group of dinner guests at a restaurant in New York when he was hideously embarrassed to find that he had left his wallet at home. Fortunately for McNamara, the *maître d'* was kind enough to let him settle his account later, instead of leading him by the nose into the kitchen to help with the washing up.

The experience gave McNamara a wonderful idea. Why not have a simpler means of payment than cash? Why not have a card that could prove who you were and that you were able to pay? Why not do without notes and coins? The result was Diners Club. Initially, consumers did not know that they needed McNamara's vision. He started in a small way with 200 members, whose cards were honoured in only two hotels and twenty-seven restaurants in the New York area. But his company persevered, and today Diners Club International is a worldwide organization.

Great companies start from small beginnings. So, apply Rule Two to your greenhouse heater. Does the idea *meet a need*? I leave it to you to work out the answer – it's time you started flying solo.

There's nothing new about what I am saying. One of my favourite historical examples of innovation is the cordon, a fruit tree which you do not have to climb to pick its crop. Possibly, cordons came from an accidental observation. But I think it far more likely that, centuries ago, a gardener sat down and asked himself: 'How can I pick all the apples without climbing the tree?' His answer was to plant trees at an angle to the ground instead of vertically. They are pruned so that they never grow above head height, and so the apples are within easy reach.

This is a really clever idea because it is not obvious that three elements combine in one approach. Reducing the height of the trees is only the first advantage. The angling also lets them be planted closer together, which replaces the crop yield lost by reducing their height, and apples which fall off have only a short distance to drop, reducing the risk of bruising.

Rule Two – *meet a need* – is also satisfied. The idea sells to anyone who is worried by heights or who wants to save time collecting the crop.

So far we have been looking at improving products. But that's just a beginning, for you must never get stuck on one line of inquiry.

Keep Looking For New Ideas

Every kind of human contact is important. Join clubs, use the old boy/old girl system, explore the networks linking you to the world. It may be a chat with friends, or a story on TV or in the newspapers that sets you thinking. Capitalize on your own interests. Mary Jacobs invented the brassière for the best of reasons – she was personally involved in the problem.

Ask yourself which groups of people you know best. These might include:

- Old people
- Children

- The sick
- Shoppers
- Travellers

Make your own list – the possibilities are almost endless, even when you limit yourself to the home. Maybe you want to help elderly people. Have you a friend or relative who has difficulty with taps or stairs? What do they want? What would they use?

Aids for the blind – white sticks, Braille, the seeing-eye dog – all had to be dreamed up by someone. There is no reason why you cannot contribute to other people's well-being. It's a noble motive, but the demand has to be there in the first place.

Here are some off-the-cuff suggestions for the infirm and for women. Apply the Rule Two test to them: Do they *meet a need*? Will they sell? You will not find 'correct' answers at the end of the following list, but you'll find a hint. In real life, you can't look up the results – you have to chance your arm, using your own skill and judgement. It's an essential part of your training to solve problems for yourself.

Drying Rail

You probably know the idea. The rail pulls out over the bath to let you drip-dry items on it. But why not also make it strong enough, solidly fixed to joists, to be a handle which helps the old and arthritic lever themselves out of the water?

Kitchen Cupboards

There's a real problem here. You cannot reach the upper shelves. Often you have to risk climbing on a stool to gain access, with a real chance that you will fall off and break a limb. It's a possibly fatal situation for the old.

Why not mount the cupboards on telescopic runners, so that they slide up and down the wall as needed? You could even have the movement powered by water pressure from the taps. Throw a lever and the cabinet descends to worktop level for easy access; reverse it, and the cupboard returns to its usual position.

But you don't even need to be that clever. The machinery could be worked by hand. Just pull the cupboard down, and shove it up again after use. Whatever you do, the runners are invisible at the back of the cupboard, and the system is cheap to install.

Dressing-Table Lights

If women make up under an ordinary light bulb, they are likely to use too much rouge or blusher. This is because the colour temperature of the bulb inclines towards the red end of the spectrum, making the redness of their make-up invisible to the eye. Once the lady emerges into the daylight, she looks slightly over the top – not a desirable result if she is a quiet personality.

Equally, if she makes up under a 'cold' fluorescent light, the results will be ideal for Ascot or a barbecue – indeed any outdoor event – but will give her a corpse-like look under normal indoor lighting.

The answer is to have switchable sets of lights over the dressing table: one set incandescent, the other fluorescent. The former give a result ideal for an evening event; the latter are just right for the day. Or, if you are a punk, you could use the idea in reverse: look like a cadaver during the day, and a recently fed vampire at night!

Which of these three ideas obey Rule Two? My view is that none of them meets a need very well, although the kitchen cupboards idea has possibilities. 'Try again' is my conclusion. But you may differ. If you do, try to explain to yourself exactly why. Write down your reasoning and come back to it in a couple of days to see if you still think it holds good.

By the way, the trial ideas can't be patented. The important thing is that they help you to test the relevant Rules for yourself.

I repeat that there are very few 'right' answers in the inventing business. I myself am far from infallible in my commercial judgements. (You'll read later about some of the mistakes I have made.)

Take skilled advice, and listen well. But you must also develop your own ways of assessing the quality of your ideas. If you decide to go ahead with one of them, the buck stops with you – no one else. You will face many questions, among them:

- Is your idea patentable?
- How do you find a manufacturer?
- Will you be competing against established interests?
- Is the law on your side?

I'll be looking at these areas in later chapters and, with the aid of my own experiences and those of fellow-inventors, will explain more of my tried-and-tested Rules for Inventors. Meanwhile, explore possibilities and markets in the ways I suggest, constantly exercising your mind. No athlete expects to break performance barriers without training. Similarly, an innovator wants to break the barriers that confine the imagination, and cannot hope to do so without practice.

CHAPTER THREE

How to Grow Rich

It's a wonderful dream: to grow successful and wealthy as a private inventor. Is it possible? Who are the winners? What do they have in common? Could you be one of them?

Private inventors are individuals who work for themselves or their own companies, sometimes with the backing of outside investors, but often on their own. Many innovators are employed by manufacturing companies in their research and development departments. It's a fascinating way to earn a living, but it does not make them private inventors. Any discoveries they make, and any resultant profits, go to their employers.

No one pretends it's easy for the individual to make a fortune, but it can certainly be done. Bill Gates, who helped to found Microsoft, the computer software company, is reported to have become a billionaire – not a mere millionaire – before his fortieth birthday.

Some writers about invention – dismal Jimmies that they are! – brush aside the existence of Bill Gates and others like him, claiming that private innovators are out of date and have little of use to offer today. These gloomsters state that our world is too complicated, and its technology too advanced, for any one person to make an important contribution to it. Private inventors, they say, do not build spaceships or find cures for cancer. They claim that major advances can now come only from teams of highly qualified experts, working with the backing of giant business corporations or national governments.

To an extent, I agree with these critics. We live in a more complicated world than a century ago, the heyday of the private inventor. Life was simpler then. Innovators who made discoveries for themselves were popular heroes, and indeed are still residents of the Hall of Fame. Who springs to mind if I ask you to name famous Victorian or Edwardian inventors? It's a fair bet that you will think immediately of at least three of the following:

- Alexander Graham Bell – the telephone
- Guglielmo Marconi – wireless
- Louis Pasteur – pasteurization
- Wilbur and Orville Wright – powered flight
- and, of course, Thomas Alva Edison

In addition to his phonograph and electric light, Edison devised a perforated film strip which helped to make the movies possible. One of his companies produced the first feature film, *The Great Train Robbery*, which led to the creation of Hollywood. Another of Edison's ideas was to make invention itself into a business. No fewer than eighty scientists and engineers slaved away at his laboratory in Menlo Park, New Jersey, and through them he owned more than 1200 patents.

Was Edison's a golden age, never to be repeated? It's true that once important ideas or products – cars, artificial fibres or whatever – become firmly established, the industries they create begin to produce their own inventions. The lone innovator is squeezed out.

But note a point which the wiseacres ignore. Changes don't have to be world-shaking miracles of engineering to be important, either to consumers or the inventors themselves.

Godtfred Kirk Christiansen saw possibilities in the toy building bricks produced by his father's company in Denmark. How could he make them more attractive to children (and to the parents who paid for them)? His answer was to invent a stud-and-tube coupling system which improved the 'clutch' characteristics of the bricks when they were pushed together. Six of his eight-stud bricks could be combined in almost 103 million ways! LEGO sprang to instant fame.

In case you ever wondered, the name comes from the Danish words 'LEg GOdt', meaning 'to play well'. The company later realized that 'LEGO' is appropriately a versatile Latin word meaning 'I select' or 'I read'. This they neatly interpreted as 'I put together'.

I don't think that any of the millions of parents and children who have been fascinated by LEGO since Christiansen devised it ever regarded it as an unimportant idea. The profits have been colossal, not only from the bricks themselves, but also from the famed LEGOLAND theme park. The company is expanding into further parks, in the UK near Windsor, and the USA.

There's a moral here. Successful innovation spins off in many different directions. It makes money in unexpected ways.

What those who are pessimistic about modern inventing also forget is that opportunities appear all the time. Change breeds change. Innovation is not like safe-breaking, with secrets being dramatically stolen from Mother Nature's jealous grasp. It more closely resembles the slow growth of a family tree down through the generations. One improvement leads to another.

Look again at our Victorian and Edwardian heroes. Bell and Marconi were themselves the heirs of Sir Charles Wheatstone, who invented the telegraph, and Samuel Morse of Morse code fame. Bell replaced the telegraph's Morse beeps with the human voice; Marconi superseded telegraphy with what was called, quite accurately, the wireless.

The next generation in this line of descent was television, which added pictures to sound. And the latest sprig on the family tree is the personal computer – the PC – which makes its users into creators and potential transmitters of all types of data – words, graphics, sound. Today's PCs are based on a machine introduced as recently as 1980 by IBM, the mainframe-computer giant. Which is exactly what the prophets of doom for private innovation would predict. Today, they repeat, the big boys run the show.

But PCs are available to any member of the human race able to plug into a reliable supply of electricity. When they were introduced, billions of people became potential buyers of innovations based on them and other microcomputers. Their appearance was a golden opportunity for private inventors.

New technologies can carry you to fortune if you climb aboard fast enough.

In the field of transport alone, the introduction of railways, iron ships, cars, and aeroplanes produced pioneers whose names sound down the years: George Stephenson, Isambard Kingdom Brunel, Henry Ford, Frank Whittle. Recently, with home computers, many more innovators besides Bill Gates have made their fortunes.

Nolan Bushnell, the founder of Atari, invented video games. His first effort, PONG, was an arcade version of table tennis. PONG was fairly basic. You turned a knob which shifted up and down a bar representing a bat, shown on a liquid crystal display. With the bat you had to stop the movement of the Pong (or white spot) from one end of the display to the other. The Pong rebounded if you succeeded in hitting it; if you missed, it vanished into an electronic limbo, and you lost a point.

I don't know how long it took Bushnell to realize the addictive power of video games, but he not only *found and met a need*, he discovered one that was totally unknown, one on which an entire industry has been built. To see how large it has grown, just look at the TV advertising campaigns mounted by the electronic games industry every Christmas.

Bushnell eventually sold out to Warner Brothers for no fewer than twenty-eight million dollars.

One of Bushnell's early employees was Steve Jobs, who moved on to found Apple, the innovative computer company. Jobs spotted the fact that many people hated using their personal computers. Why? Because the opening display on computer screens offered only a symbol like this, 'A>' – the notorious 'A' prompt. It was totally unhelpful. If you weren't trained in computer programming, what were you supposed to do next?

Jobs's Apple Macintosh computer offered a new approach. It put advice and instructions on screen in picture form. For instance, if you wanted to delete a file (computer-speak for to wipe out something you have written), you consigned it to a picture of a wastebasket. The Mac program became famous for its user-friendliness.

It was a typical example of the value of both Rule One and Rule Two. Millions of Apple Mac sales prove that Steve Jobs both *identified the problem* and *met a need*.

Alan Sugar is not an inventor, but he is certainly an innovator. His genius, like that of Steve Jobs, was to understand that computer technology can get in the way of people who want to use it. He asked himself how computer innocents could find their way through the fog of industry jargon which baffled everyone except the experts. How could users sensibly choose a microcomputer for their needs, select the right operating system to run on it, and pick a suitable printer? There were too many choices and not enough information, when all that most customers wanted to do was turn out high-quality letters and documents.

Sugar dreamed up the Amstrad PCW: a package of monitor, keyboard, printer and operating system, supplied in one box. Plug it in, switch on, and away you went. His customers no longer had to make complicated decisions about dozens of types of equipment, choices which could turn out to be expensively wrong. His PCW's price was low because it used a mature – i.e. discarded but still workable – technology, and the whole product was marketed as an advance on the typewriter. He *identified the problem* and *met a need* that no one else was meeting. Sales of over a million Amstrads prove that he was right.

You see the possibilities? Private innovators *can* be successful today. Famous, too, if recognition is what you're after. You may have come across some of these inventors' names:

- Felix Wankel – the rotary car engine
- Sir Clive Sinclair – the pocket calculator and digital watch
- Edwin Land – the Polaroid instant camera
- Ernó Rubik – the puzzle cube
- Ken Wood – the food processor
- Robert Moog – the music synthesizer
- Ray Dolby – sound systems for magnetic-tape recording

What personal qualities do *you* need to see your name inscribed in the Hall of Fame? There's no general recipe, but successful innovators do have certain characteristics in common. The first is *firmness of purpose*.

Alexander Moulton, an expert in suspension systems, designed a bicycle claimed as the first major change in bike design since the linked chain was patented in 1896.

The Moulton bike's most obvious feature was its 14-inch

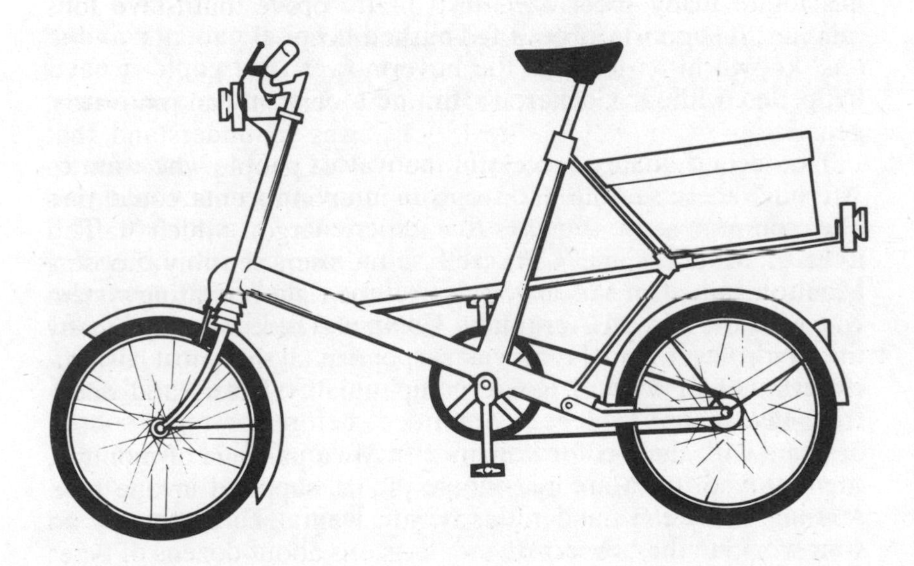

wheels. They lowered the frame nearer the ground, making it more stable, easier to manoeuvre and handy for carrying loads. Unique sprung forks ironed out shocks created by the stiffness of the small wheels.

No manufacturer would look at the new bike, so – firm of purpose – Moulton founded his own company to make it. His zippy little machines became a symbol of the 1960s, along with the Mini car and the miniskirt.

Christopher Cockerell wanted to make boats go faster. He tried no fewer than four designs in his attempt to find how their hulls could be supported on a cushion of air. His final, balsa-wood, model travelled at 13mph over both land and water. But when he approached manufacturers with his prototype, he discovered that boatbuilders thought his idea was an aircraft, while plane makers said it was a ship!

Next the Ministry of Supply classified Cockerell's vehicle as 'secret', delaying the project until the National Research Development Corporation sponsored a seagoing hovercraft. Cockerell worried that hovercraft might dive into the water if they were driven fast across waves. He decided to give them flexible skirts – and was ridiculed because no one believed that rubberized fabric could help support a heavy vessel.

But the inventor won through, in the end. His hover principle has found many specialized uses in transport, both civil and military; it supports airbeds for burned hospital patients; and is best known in its form as the hovermower. But could it have happened without Cockerell's firmness of purpose over many years?

The second quality successful innovators share is *experience*. Although there's nothing to stop you innovating in any field you like, common sense suggests that experience is important. The lives of Moulton and Cockerell show marked resemblances. Moulton gained an MA at King's College, Cambridge, spent the war with the Bristol Aeroplane Company and later developed the Hydrolastic and Hydrogas suspension systems for BLMC (forerunner of today's Rover company). Cockerell graduated from Peterhouse College, Cambridge, before serving a couple of years with the Bedford engineering firm of W.H. Allen. He then moved to Marconi, where he filed thirty-six patents relating to wireless and radar before leaving the company to start work on the hovercraft.

The similarities in their careers are obvious. It's the same story in Continental Europe. Dr Glauco Corbellini, from an ironmaster's family, qualified as an engineer in Italy before making no fewer than forty improvements to iron-making. An interest in air-flow engineering led him to invent a slatted sail for ships. Watching the sails clicking in the wind gave him an idea for flip-over signs to supply large-scale travel information. You've seen them at airports and railway stations. The 1045 for Orlando departs from Gate 12. Flip, flip, flip. The 1900 for Liverpool leaves from Platform 6.

Dr Giesel von Gieslingen from Graz, in Austria, trained as a locomotive engineer. His background helped him invent a new exhaust blast for steam engines. It increased their thermal efficiency from seven to a phenomenal eighteen per cent. Von Gieslingen's flat blast injector kept Austrian steam locomotives going for at least ten years past the time when they should have been pensioned off in favour of diesels.

The common denominators are that Moulton, Cockerell, Corbellini and von Gieslingen trained as engineers, were vastly experienced by the time they produced their most notable ideas, and had track records of innovation. If you quail because you fail this experience test, there is no need to despair. The solution is to become your own expert.

Dr Rudolph (Rudi) Breu started life as a guitarist. He taught himself chemistry in Munich with money he made from playing the guitar in nightclubs. Breu invented a foam protective paint which is used to proof surfaces against fire. He took his idea to the German company BASF, which set him up with a company to produce his intumescent paint, as it's known. The innovation not only made money, but also gave Breu a career – not to mention the enjoyment of continuing his research at company expense!

Another solution is to hire the help of other experts (as Alan Sugar did with his Amstrad PCW).

J.C. 'Dai' Davies was no specialist – he sold advertising space – but he noticed how letters slip about on children's water-slide transfers and used his observation to create a wet-lettering system for graphic designers. It was successful, but messy. Davies needed expert help. It came from Frederick Mackenzie, a printing consultant, who found an elegant way to make dry transfers. He placed a letter (carried on a master sheet) in a precise position on the page, then used the pressure of a stylus or ball-point pen to transfer the letter exactly where he wanted it to be.

Mackenzie had invented Letraset. It gave advertising designers hands-on control over what they wanted to do, and changed the face of display work for ever.

The third quality great innovators share is the ability to *think like a child*. Letraset evolved from watching a toy. Practise examining life with the innocent eye of a child but the knowledge of an adult. Children see problems clearly, without the handicap of preconceptions. Nothing muddles their view.

My own first invention, if you can call it that, came when I was nine. I had a school chum who was hard of hearing, but who wanted to listen to gramophone records. I observed – as many grown-ups had already done, but I didn't know it then – that when you twang a string between your teeth, you can hear the sound, even if you plug your ears.

The conductivity of sound through bone is well known, but I am going back to a child's process of discovery. I wondered what would happen if I made a plectrum – flexible celluloid on a piece of balsa-wood – and pushed a gramophone needle through its other end. I got my friend to hold the plectrum in his teeth, with the needle positioned on the revolving gramophone disc. He tracked the groove across the disc as it spun. It's not easy to listen to a record that way yet, deaf as he was, he could hear the music!

Another childhood story. My godson was watching his father scratching dead leaves from a large strawberry patch at the end of the growing season. Done by hand, it is a tedious and time-consuming job. Richard observed the performance with interest before remarking, 'Daddy, why not use that rake with ten prongs in the garage? You could scrape up the leaves faster without damaging the plants.' His father (a first-class Honours graduate) looked at him. Four-year-old Richard was right.

Sir Barnes Wallis, shrewd inventor of the Dambusters' bouncing bomb and designer of airships and aircraft, made the childish test question 'Why not? . . .' his touchstone for generating ideas. Try it. It works.

German chemist, Dr Dietrich, was looking for a less messy way of applying glue when he noticed his lady assistant making herself up with a lipstick, pushed out of a tube. 'Why not? . . .' he asked. He came up with PrittStick, glue paste in a cylinder, advertised as 'the non-sticky sticky stuff that sticks paper and card'. The UK Patent Office estimates PrittStick's worldwide sales for the Henkel company at over £140,000,000 to date.

Swiss engineer Georges de Mestral repeated an experiment many children perform at school. He peered at burdock seeds under the microscope after they stuck to his trousers. Thinking like a child while benefiting from adult knowledge, de Mestral saw how the burdock's tiny hooks made its seeds cling to the hairs on animals' coats and to human clothing. 'Why not? . . .' he pondered. Eight years later, his answer was Velcro (from the French words '*velours*' (velvet) and '*crochet*' (hook). The final burdock-like product was made of two nylon strips – one with thousands of small hooks, the other with tiny loops. The Patent Office says the idea is already worth £5,000,000 in the UK alone.

Budding innovators quickly run into clever persons who are only too keen to assure them that no private inventor can achieve fame and fortune today. The stories in this chapter prove they're wrong, and are intended to boost your confidence. To be realistic, the odds are indeed against your becoming wealthy. Nobody says that the path of the innovator is easy: many inventions get nowhere and many innovators give up. But: fated to fail? Never!

CHAPTER FOUR

Inventors Today

As President of the Institute of Patentees and Inventors (IPI) I am in touch with hundreds of private inventors, and get involved, if at second hand, in their many struggles. Few of them are outright winners of the kind we looked at last chapter. Some work at inventing full time, many more have bright ideas which they try to sell while they hold down full-time jobs. They are teachers, secretaries, lorry drivers, average members of the human race – but with a spark in their eye. Many will get nowhere, but among them are individuals who may achieve fortune in the years ahead.

I've surveyed a selection of IPI's members to find out who today's innovators are and how they think. Among the questions I asked were:

- What qualifications do they have?
- What have they invented?
- How successful have their inventions been?
- How are women inventors getting on?
- What are the problems of innovators today?

Let's look at some of my findings.

Do Innovators Have To Be Engineers Or Scientists?

No. Only a quarter describe themselves that way. Another twenty per cent run companies or are self-employed, often because they are in a business which is connected with their inventions; many of these have technical qualifications. The medical profession makes up a further six per cent.

The remaining forty-nine per cent – nearly one in two – come from a joyously mixed list. They include an advertising man, a sculptor, a shorthand typist, a practitioner of natural healing, a musician, an average adjuster, a gardener, a weighbridge operator, and an owner of holiday flats. Surprisingly, retired people, who might be expected to have enough spare time to pursue innovative ideas, amount to only five per cent of the total.

What Do Private Innovators Invent?

You might believe that individuals and members of small companies would be chiefly involved with innovation in the home. It's not true. Nearly forty per cent of their inventions are industrial and commercial, twelve per cent are concerned with transport, and eight per cent with agriculture and horticulture. Ideas connected with the home, plus shopping and DIY, add up to only twelve per cent.

Included in the survey were some of the inventions I have presented in the Patent Pending section of the BBC's *Tomorrow's World* programme. These are generally of a high standard, though they may not have found a manufacturer or won commercial acceptance at the time when they were shown on television. I *know* that they are good because I have studied them myself in depth.

Patent Pending innovators are slightly more engineering-inclined than the much larger number of people on the membership roll of the IPI. But not a lot: they include a farmer, a furniture maker, a bank official and an employee of Youth Concern. The types of inventions they produce also match the IPI pattern. Let's look at a few of them.

- Leslie Joseph's aid for dinghy sailors and yachtsmen, Buoy Grab, helps them moor safely and quickly. The device is pushed through the ring at the top of the buoy and locks on to it, threading the rope through in one movement. It is a one-handed operation which makes it unnecessary to lean over the side of the boat.

- Stephen Harding invented a pneumatically controlled ten-speed internal gear hub for bicycles. You squeeze a ball on the handlebars to go up a gear, another to go down a gear. The air pulse from the ball trips a latch which releases a pin. There are no control wires – wheel motion does all the gear-changing work.

- Gerry Lloyd improved the supermarket trolley. You know the problem: trolleys and airport baggage carts seem willing to move in any direction but the one you want them to. Gerry added a fifth wheel which gives complete manoeuvrability but cuts out sideways drift (see photo section).

- Colin Branston's Fencability is an adjustable electric fencing system for farms. The farmer can move the fence lines around without switching off the current – very useful if he has stock at heel. Fencability is self-tensioning.

This partial list suggests that private inventors do have sensible, serious ideas. They are not just making widgets or tiny improvements to existing products.

Do Inventors Work Only In Their Own Areas Of Knowledge?

No. But there's no doubt that this is a sensible way to operate. For example, if we look at IPI members with medical expertise, we find London surgeon Riad Roomi and his Ready-Stitch.

Case history

Mr Roomi worked in a hospital casualty department where he regularly found himself stitching up horrific wounds sustained in accidents. He found it distressing to put stitches into skin when the patient was not under a general anaesthetic or was a young child. He asked himself Barnes Wallis's question: 'Why Not?' In this case, why not find a new method of closing up the skin? Why not stitch the surgical plaster instead of the wound?

Ready-Stitch was the result. It comes in several versions, depending on the type of operation involved. The simplest is plasters which are stuck down on the opposite sides of the wound. Threads linking the plasters are then tightened, pulling the flesh together.

No needles are used, and, because the edges of the wound are not frayed or damaged by being punctured, the results are better cosmetically. Ready-Stitch marks the end of the appendix scar that mars the beauty of bikini wearers. Among its other advantages are that it is painless – valuable when treating children – quick and easy to apply, and it gets rid of needle-prick injuries, eliminating the risk of AIDS and hepatitis-B to doctors and nurses. It saves time in busy accident departments and cuts down the use of local anaesthetics.

Mr Roomi's first experiments – which began with making a working model of a wound – were pretty low-tech: 'I made a slit in the surface of my rubber living-room sofa, to reveal the foam sponge underneath. It was a simulation of damage to the human skin.' The sofa's wound responded successfully to treatment,

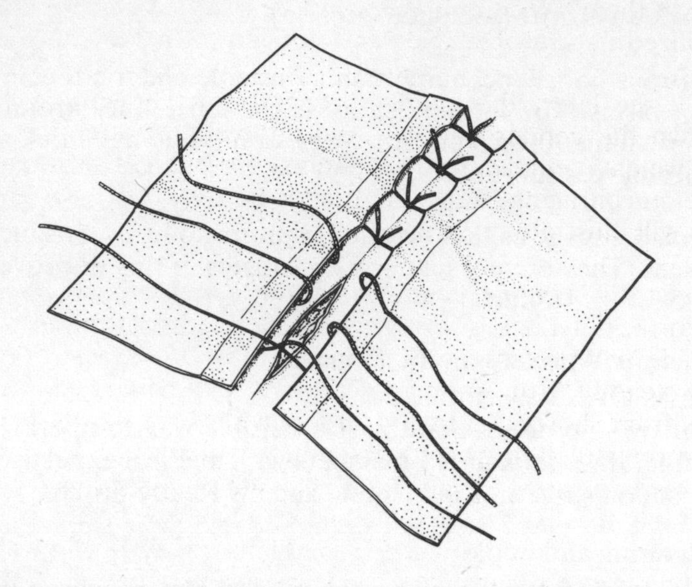

and Mr Roomi moved on to try his idea on live patients. When their wounds healed as well as with the traditional method, but without scarring, he knew he had a workable idea.

The medical profession, and judges at contests for innovation awards, agreed with the innovator. He was flooded with inquiries from doctors and won first prize in the Toshiba Year of Invention competition. In the Department of Trade and Industry's SMART contest, he was a Stage 1 winner, receiving an award of nearly £34,000. I am proud to say that the IPI recognized his good work early on and presented him with the Richardson Gold Medal, which is awarded annually for the invention that best anticipates future needs or meets the known demands of society.

Ready-Stitch is a brilliant invention by a specialist working in his own field. But must innovators remain within their own areas of expertise? It isn't necessarily so, and the experience of women inventors proves it.

How Many Women Inventors Are There?

A mere five per cent, I am sorry to say – nowhere near enough. But I live in hope that the new age of liberated women will change the balance. After all, I argue that innovation can be self-taught. Who's to deny that women can train themselves as well as men? Who's to say that women don't have the flame of creativity in them? Wouldn't the world be a better place if the more civilized nature of women was applied to improving our future?

As we'll see later, the younger generation is beginning to break down the gender barrier. And although we are pretty short of weighty engineering inventions from adult women, there are encouraging signs on the domestic front. A few examples will illustrate my point.

• Architect Vivi Hagrup came up with a lockable horse tack carrier, Portack-et-Lock. (Now here's someone inventing in an area outside her professional speciality.) Thefts of tack have increased steadily, from private tackrooms and riding schools, as well as from shows and livery yards. Saddles are expensive. They are even stolen in transit, while the horse and its tack are on their way to a show.

To beat the thieves, Vivi designed a locking device based on a plate with a ring and hook, which is held in position on the wall by special screws. A security chain threads through the ring. It is then padlocked on to a specially shaped carrier which is wrapped around the saddle. Two wallplates are required, one for the tackroom and the other for the trailer or horsebox. Portack-et-Lock is easily transferable between the two – hence its name (see photo section).

• Josephina Banner is a professional sculptor and designer who is interested in aids for the disabled. Her work in this field has nothing to do with sculpture, and again the message is clear: innovate where your interest takes you; there is no rule.

Josephina feels that her best idea is a patented touch-carpet for blind children, which she says 'was approved by all the top shops and authorities'. The carpet is woven in a varied natural pattern which lets unsighted people feel exactly where they are standing on it; it tells them their position, which makes it ideal for theatrical and dance performances on an open stage.

• Mrs Patricia Van Emst invented an absorbent pad to prevent pressure sores. The retired nurse wanted to prevent sores appearing on the bodies of long-stay hospital patients and home-care patients suffering from incontinence. Her pad, which is disposable and incorporates bubble plastic dressings for hospital use, can also be adapted for packaging soft fruit or other delicate articles.

All this is admirable, and I eagerly await the day when a private invention by a woman rivals the hovercraft or pocket calculator in worldwide impact.

Can You Make A Living As A Professional Innovator?

We're not looking here for the big pay-out, the jackpot winner that means you never have to work again. No one can safely make predictions of that kind. We're considering the possibility of turning out a steady flow of ideas and inventions, getting them manufactured and sold, and earning a regular living. There's no doubt that the best way to achieve this is to found your own company, manufacture and maybe distribute your own product, and make sure that you keep enough spare time to carry on inventing.

Case history

Frank Arthur is chairman and chief executive of the Envopak Group. Envopak is one of those ridiculously simple ideas that makes you gnash your teeth because someone else thought of it first.

Many large organizations, from the government through manufacturing companies to retail chains, send huge quantities of routine paperwork between their head offices and branches, shops and depots every day. Each piece of correspondence needs a separate envelope, separately addressed and separately stamped or franked. Each piece may cause a security problem if it goes astray, and with thousands of documents involved, every so often one of them does go missing.

Answer: put all the mail for one address in one Envopak, a bag capable of holding 2000 expendable envelopes. It has just one removable label which needs stamping or franking only once. The Envopak itself is sealed for security. The company claims a minimum working life of five years for the bags, which come in a wide range of sizes (see photo section).

Frank Arthur has taken advantage of his success with Envopak to patent a new device: the Posicheck Seal, which protects the seals on warehouse doors and keeps containers secure during transit. Establishing if thieves have broken into a large storage building or a container is often difficult if the villains know their business. Padlocks can have their keys stolen or duplicated. Posicheck gives a visible indication if tampering has taken place. Envopak claims that it's impossible to open and replace without clear evidence that something has happened to it.

'From the original Envopak, numerous advances and other

security product innovation have stemmed,' says Frank Arthur. 'The company directly now employs some 400 people. Last year we exported our various products to eighty-three countries.'

Now *that* is what I meant at the beginning of the book when I suggested that we should concentrate more on innovating because we are good at it!

I received many replies in the survey from businessmen in a similar position to Frank Arthur. Their business supports the innovating and when all goes well, innovating supports the business.

Let's look at one more example where, although the relationship between its head and innovation is much the same, the nature of the company is very different from Envopak.

Case history

Tom Ketteringham is chairman and managing director of Martek, a firm based in Redruth which runs a special line in DIY, tradesmen's and professional equipment. He holds the patent for a drill-bit sharpener. The Professional is powered by an ordinary handyman's drill and comes in several designs. It sharpens steel or masonry drill-bits between 2mm and 12.5mm, and offers variable cutting angles to suit all drilling applications and a stone-dressing attachment. Martek sold 35,000 sharpeners in the first year and a massive 160,000 in the fourth.

Tom Ketteringham confirms my analysis when he says: 'Our first product required a number of small investors. All other products have been financed from sales of the first product.'

Clearly, what we might call Route One to earning a living as an inventor is to run your own company, inventing as you go and selling what you invent. But there are other routes.

Case history

Head of his own company, Design Automation Ltd of Maidenhead, Bob Redding has been inventing for fifty years, mainly as an independent problem-solver. In the 1950s he found instrument developments in the petrochemical industry were leading to an intractable problem: the safety of electrics in flammable atmospheres, where sparks could cause fires and explosions.

Bob Redding evolved a type of safety barrier which made circuits intrinsically safe. His idea, using zener diodes, was simple and easily understood but difficult to put into action because of regulations and commercial interests. While promoting the zener barrier he became an independent consultant, and his promotion of the idea in industry and trade associations led to the barrier being recognized as a basic solution, now known and used worldwide. It's an encouraging thought for up-and-coming innovators that Bob Redding found consultancy, lecturing, writing books and following up further ideas paid more than conventional manufacturing.

Then came another discovery: how to avoid gas explosions in the home. In 1985 British Gas ran a project to develop the 'Gas Meter of the 21st Century'. On behalf of a large international organization, Bob Redding put in a proposal based on a dedicated electronic chip and ultrasonic transducers. You can catch the way his mind works from some thoughts he voiced on this project.

'The purpose of the gas meter is the safe distribution of gas, not just accountancy. The intelligence and monitoring capabilities of an electronic system should be exploited to give complete safety monitoring of the entire pipework. Further, the polythene wall of the gas pipes now used can carry strip-line "electrical circuits". These can not only power the gas meter, but also provide a high-speed communication network for metering other domestic supplies like water and electricity, as well as the gas itself, telephone and cable TV services.'

Bob Redding has patents for the strip-lines, modems and meters required for these proposals. This is a man who can think big, as well as look after the minute details which must be attended to if an invention is to succeed.

Did you notice the crucial part of my sentence which talked about British Gas? 'On behalf of a large international organization, Bob Redding put in a proposal . . .' He does not invent 'on spec', or if he does, he soon finds a client to work for. Nor is he distributing the products he designs, as Frank Arthur and Tom Ketteringham do. He makes his living from innovating and invention for others. You might call this Route Two for earning your keep: as an independent.

Bob Redding adds a cautionary note of his own: 'Be warned: you need a number of balls in the air – and most of them will fall. The trick is to get some return before they do!'

Route Three is followed by only a handful of people.

Case history

Alan Wilcher invents, but only when he has to. He prefers to be a midwife to other people's innovation, coaxing the new-born invention successfully from the womb into the cold world. If requested, he'll be nanny to the invention, schoolteacher, drill sergeant – and marriage maker as well – at various stages of its life. That is to say, he will improve the concept, help refine a prototype, advise on pilot manufacture, find a licensee or approach distributors with a well-prepared sales proposition. Alan Wilcher is an innovation consultant. Or, as the name of his company suggests, an 'Imagineer'.

Much of Alan Wilcher's work is commissioned by companies which want their own R and D efforts supplemented by an outside expert whose professionalism they trust. He himself is an agricultural and road transport engineer with international management experience. From its base at Ramsey in Cambridgeshire, Imagineering has handled over 100 projects since he founded the company in 1988. Some of these have been for governments and some for giant corporations, but others were for private inventors who persuaded Alan Wilcher that he should take them seriously (he will courteously decline to become involved if he thinks an invention is a no-hoper).

Alan Wilcher is also determined to promote his clients rather than himself, so, although his company is mentioned only briefly in the next two case histories, you must visualize Imagineering behind the scenes, guiding and advising along the way.

Case history

The idea of Logmaster began life when Jim Campbell, a farmer from Harrogate, had to cut up over fifty tons of trees stricken with Dutch elm disease. He found that the cambium, the outer

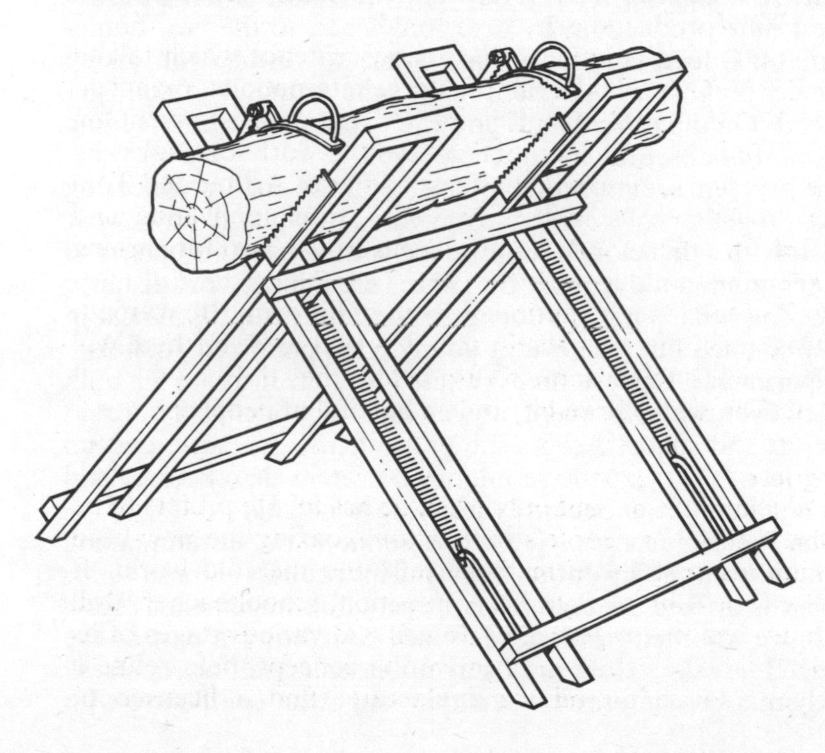

part of the trunk which normally keeps the tree alive, had rotted and separated from the bark, which slithered and rolled around in a most alarming way when the saw chewed into it.

He was well aware that cutting up any wood with a chainsaw can be very dangerous, especially where tree limbs and mixed-growth timber are involved.

After Jim Campbell had designed a first solution to the problem, he hired Imagineering to help him convert it into reality. They helped him to produce models for both professional and domestic markets to go into production.

In operation, Logmaster is a joy to watch. The operator simply throws a log into the crutch of an X-bar trestle the size of a normal sawbench, and two spring-loaded jaws snap over to fix it in place, ready for sawing.

Case history

Another farming invention – it must be Alan Wilcher's Somerset origins calling to him – which has involved Imagineering is an ingenious feeding system for dairy cattle. Yoke-l grew from an idea put forward by Joe Dyke, who lives near Devizes.

Stock work is a scientific art these days, the aim being to increase milk yields while spending as little as possible on expensive bought-in feed. You want the better cows, with the highest milk production, to have free access to the best home-grown bulk feed, but you don't want to disturb the group dynamics of the herd by upsetting the other cattle, who don't get the feed. On the other hand, you can't split the herd at feeding time *without* upsetting it.

The problem seems insoluble, but Joe Dyke and Imagineering worked together to provide the answer. The 'special' cows wear a collar with a detector. It lets marked cattle through a barrier to the high-grade food trough; the others are kept a short distance away. The herd is not split because the cattle are still almost in bodily contact, but feed is now going to the preferred animals.

Experiments suggest that Yoke-l gives an increase in milk yield of over twelve per cent, as well as an eight per cent boost in butterfat and solids. Across the entire herd, the cattle end up eating less of the expensive imported protein than before, and their health is better because there is less jostling.

Alan Wilcher charges a fee for his services, rather than taking a royalty from the invention's sales. This probably works to his loss but is a fixed point in his contractual arrangements. 'How can I give my clients independent advice if I own a part of the project?' he asks. 'How can I stay unbiased myself or represent the client's best interests when they might be conflicting with mine? I don't want to find myself saying let's go ahead, because I need the money, when the client isn't ready.'

There is no way to estimate the fees for a project in advance – everything that Imagineering does is a one-off – but typical costs range from £2000 to £5000.

And that is Route Three to becoming an independent inventor – one who relies on no other employment.

Overall, unless you have good reason to be very confident – good reason being nothing less than profit already in your hand – my advice is: 'Don't give up the day job.'

And there we have an overall portrait of inventors in the UK today. We haven't yet reached their opinions – we'll come to them in later chapters on patents and troubleshooting. Nor have we considered young people. They're a group of innovators that my survey couldn't pick up, as they are too new to inventing to have joined the Institute.

How easily children invent! Their ideas are usually free of the influence of authority figures telling them what is possible or impossible. They haven't yet acquired those fixed notions which can block original thinking.

Children see reality with tremendous clarity when they are very young, but their constructive period begins at about thirteen. At this stage, their inventions are usually simple, often already familiar ideas. But sometimes they are outstanding. Here is a selection from the Young Engineers for Britain contest.

- Twelve-year-old Ross Carden from Sutherland noticed that a fire could be caused if a switched-on iron, standing on its element, falls over. His safety iron has two micro-switches built into its circuitry. When the iron is standing up, the switch on the base is activated and allows the iron to heat up. The second switch, on the handle, must be pressed by the user when beginning to iron. If the iron is left standing up and then falls over, the switch on the base opens and turns off the current. 'This could be useful to older people,' said Ross, 'and to young mothers whose attention is distracted due to young children, or animals accidentally knocking the ironing board.'

- James Higson set out to improve a situation every motorist knows: when automatic signs at car parks show 'Spaces' when you join the ingoing queue, but switch to 'Full' while you are waiting and unable to go elsewhere. James's SpaceFinder monitors every place in the park with an infrared monitoring system which detects the presence of a parked car. Data from the sensors is fed to a central processing unit.

'The number of spaces left and the identity of floors which are already full are displayed on large, easy-to-read digital signs at the entrance to the car park,' James explained. 'At the entrance to each floor, the location of every available space is displayed, enabling the motorist to drive straight to an empty bay.'

When I first came across SpaceFinder, I was particularly impressed by James's attention to the commercial side of its development. He shrewdly pointed out that it would give car-park owners who invested in it a significant advantage over competitors; it was the SpaceFinder parks which would be full, earning maximum revenue for their owners. If only more inventors thought this way, about the moneymaking side of their ideas, they would immensely improve their chance of selling their inventions.

- A device that really stands out in my memory is a physiotherapy aid for stroke patients. It was devised by Samantha Haines as her electronics A-level project while she was in the sixth form at Cheltenham (boys!) College. Samantha got the idea during a week's work experience in a hospital physiotherapy department. She found that stroke patients often have difficulty in regaining their sense of balance, yet there didn't seem to be a suitable piece of equipment available to help them. 'Why not?' she asked herself, in the best Barnes Wallis manner. But she didn't immediately rush off to design the machine. Instead, she asked around to find out exactly what the market required. Sensible girl. If only more inventors did the same.

With the results of her research, and advice from her electronics masters, Samantha then designed the circuitry for two platforms which the patient stands on, one foot on each. The results, feeding to a visual display, show how the patient's weight is distributed. The patient watches the display and shifts his or her weight to come to a perfectly balanced position. This is difficult if you've had a stroke, but the device allows patients to manage it in their own time without constant attention from a physiotherapist.

All this is easy to describe, but how was Samantha to measure weight? A mechanical system would have so much movement in it that the patient's balance would be more disturbed than ever. Load cells were too expensive. The obvious answer was specially adapted strain gauges, but where could Samantha get them from? After she made a presentation to the manager of the development department to show him that her idea was workable, Dowty Aerospace Propellers gave her the help she needed.

Samantha adopted the same professional attitude to all aspects of her innovation: choice of materials, simplifying the design, getting help where needed, even future marketing. The result was that she won the £1000 award for the best Young Engineers for Britain electronics project – and she thoroughly deserved it. She has set up a company to manufacture an improved version of the device, which she intends to sell at a price that no physiotherapy unit can afford to turn down.

I have talked at considerable length about Samantha's invention because she and her teachers realized that an innovatory idea is not enough in itself. Identifying a need, tailoring your product to users' requirements, hitting the right price, using the correct materials and considering safety aspects are all vital to success. Just one criticism: I wish the aid had a name! No doubt it soon will, if Samantha's track record is anything to go by.

Samantha Haines is a forerunner of a new wave of female innovators. Entries by girls for the Young Engineers for Britain contest are running at one third of the total, and rising. Premier awards fall to them with increasing regularity – so much so that there has been serious discussion about whether girls need their own special award. Why not let them compete openly against the boys, and leave it at that? I think this is evidence that the playing field between the sexes is becoming more level.

Other innovation schemes are increasingly dominated by young women. Shell UK's STEP programme places undergraduates in industry over the summer vacation to carry out special projects, and the ladies are doing well (see page 215).

For those older than twenty-five, whether male or female, the message is: keep on trying. It's the way you view life that matters, so learn to *think like a child*.

CHAPTER FIVE

How Things Go Wrong – Part One

A useful mental trick when you are wrestling with an inventing problem is to stop worrying about the winners you are attempting to outdo. Look instead at the failures. You may learn more from analysing the reasons why they went wrong than you will from the people who got it right. Three stories from television and films are enough to show what I mean.

Was The Innovation Overtaken By A Superior Product?

John Logie Baird produced television pictures using a mechanical scanning device known as a Nipkow disc. Results were coarse-grain and crude, but Baird steadily improved them to the point where the BBC could transmit experimental broadcasts. A year later Baird was selling his Televisor, the first mass-market transmitter. His prospects looked good. But by 1936, when the BBC started broadcasting regularly from Alexandra Palace, it had a choice between Baird pictures containing 240 lines per second and rival Marconi-EMI pictures with 405 lines, nearly twice the definition. The latter were totally electrical in origin, based on the cathode-ray tube.

Baird's system was dropped. Despite the fact that he was the pioneer, that he had been developing television since 1922, that he had immeasurably improved his pictures since their early days, the moment a better product appeared, he was finished. It doesn't matter what your track record may be – the market is interested only in the future.

Was The Product Offered In The Wrong Place?

Smells pumped into cinemas to accentuate the impact of exciting scenes on the screen failed commercially. But travel through the Jorvik Centre at York, and your understanding of life in Viking times is reinforced by subtly sprayed odours of tanning, cooking and less enjoyable things. The sniff technology which did not work with feature films has found a niche in themed displays. Note that the product was not only misplaced in the cinemas, but was at the wrong time. It had to wait twenty-five years for the rise of the thematic museum to find its own moment in history.

Was The Product Beaten By Superior Marketing?

Nobody who knows about home video-recording disagrees that Betamax is technically superior to the VHS system. It was first in the field, as well. But VHS came from behind to take command of the market and Betamax has largely been relegated to educational uses. It's not enough to have the best product if you don't sell it efficiently.

Checking against failures can reveal hidden truths. Now turn around the questions I asked. They are mirror-image to three problems which arise all the time:

- Is the innovation good enough to beat the competition?
- Is the innovation being sold in the right place?
- Is it being marketed properly?

These are questions which can only be answered when the details of each problem are spelled out, but remember that they cannot be answered at all if they are never asked. This point leads me to the next of my Rules for Inventors:

Rule Three: Keep on learning

Question, question, always question. Making false assumptions about an innovation leads to disaster. Let's stay with failures to see what further lessons – I have identified five – we can learn from them. When we look at some of the odder examples, one message becomes immediately apparent. Some patents are not simply trivial – they are totally without merit. Their only benefit is that they teach you what *not* to attempt when patenting.

Lesson One: Make Sure The Invention Works

Many examples of failure to meet fitness for purpose come from the period before the First World War, the heyday of the private inventor. But a much earlier specimen is a contender for the title 'Silliest Patent of All Time'.

- In 1718 James Puckle, Gent – many calling themselves

gentleman became patentees; perhaps they had nothing better to do with their time – patented a repeating gun to repel pirates attempting to board ship. He obviously had in mind the Arab and Turkish corsairs who were then the plague of the Mediterranean, but he was open-minded enough to recognize that descendants of the Crusaders could turn rogue as well. His gun, therefore, fired round bullets at Christians and square ones at Muslims.

• Another weapon – warfare always has its black humour – is the headpiece with a gun concealed in it, Pratt's hat. This was securely tied to its wearer's head by a chinstrap, and when the wearer blew into a tube inserted into his mouth, he operated a device above, which squeezed the trigger and fired the gun.

Dating from 1917, Albert Pratt's helmet was no doubt inspired by the carnage of the Great War, although he did not realize that he might add to it in a massive way – by killing off men on his own side. Anyone who has ever suffered the kickback of an ill-held rifle will flinch at the thought of the number of broken necks Mr Pratt's helmet-gun could have created.

• On the home front, an American lavatory seat from 1869 taxes the imagination. It required four rollers to be mounted round the top edge of the bowl, each being slightly hollowed towards its centre, which rendered it 'impossible for the user to stand upon the privy seat'. The rollers 'in the event of an attempt to stand upon them, will revolve and precipitate the user on to the floor'. But what was the problem with people standing on the seat that the inventor was so keen to solve? Peering into the next cubicle? Perish the thought.

• However, the overall winner comes from that home of lost ingenuity, aeronautics. In 1886 Charles Wulff of Paris patented a balloon propelled and guided by vultures. Thoughtfully, he conceded the possibility that eagles or condors would do equally well. Heath Robinson could not have come up with a better balloon than Monsieur Wulff. Underneath it was a car containing the passenger. He gave orders through a speaking tube to another person (call him the flight engineer) who stood in a second car mounted *above* the balloon. The vultures (or eagles or condors) were strapped to this same platform.

Above the whole insane contraption was a permanently open parachute, topped by a gay medieval banner floating in the breeze. The chute presumably acted as a parasol to keep the sun off the birds. By turning a handwheel, the flight engineer could make the condors (or vultures or eagles) fly level in any direction. Using a roller mechanism, he could direct them up or down.

Wulff's concept has a truly *Alice in Wonderland* vigour. Apart from any little difficulties associated with the amount of bird power generated, or the period of time for which the creatures (which are basically soaring birds anyway) could continue to fly level, there is a problem which Wulff himself admits, though he hardly addressed a solution. 'It may be observed,' he says, 'that the birds have only to fly, the direction of their flight being changed by the conductor quite independently of their own will.' Just so. The birds have only to fly.

These are ideas that reinforce the popular image of the lunatic deviser, cackling maniacally to himself in his workshop at the top of an ivory tower. Or sticking straws into his hair to ventilate his brain while he boils pots of glue in his garden shed. Many diverting books have been written on this theme, not least for children. All the way from rompers to adulthood, we are spoonfed the concept of the inventor as a being who stands at a slight angle to reality.

With the benefit of twenty-twenty hindsight, it is easy to make fun of some inventions. But hindsight also distorts, for we forget the circumstances which caused them to come into existence. Take the spade with a counter built into its handle which recorded how many times the spade lifted a spit of earth during the day's digging. A ludicrous idea today, when we all do our own gardening; a prize piece of useless information. Yet it was not so idiotic in the days when many people employed gardeners: the invention told a cost-conscious landowner whether his employees were working or not.

Equally, Charles Wulff was addressing a genuine problem with his vulture-driven balloon. It had been proved that making airships cigar-shaped would solve the chief problem with spherical balloons, which was that they could not be steered. But steering gear was still not perfected. Aerial oars and ships' screws had been tried and found wanting. Motors to drive the screws were heavy and difficult to operate. Wulff therefore identified a real problem. So did Puckle with his gun (piracy) and Pratt with his hat (self-defence). What they did wrong was to come up with unworkable solutions.

LESSON TWO: MAKE SURE THE IDEA IS THOUGHT THROUGH

Do not believe that oddball invention ceased with the Victorians.

- In 1955 James Greco had a really appealing idea, spoons and forks whose handles were hollow and punctuated with holes so that they could be played as whistles. You can imagine Mr Greco getting the thought while watching his children or grandchildren messing around with their food. But what happens when the children start playing tunes while food is still on the spoons is awful to contemplate.

- Harold Dahly's solar hat of 1967 suffers from a similar failure to follow the idea through. A solar cell in its crown provides power to run a motor which drives a fan inside the hat. Cooling air is drawn in through holes in the side of the headgear, and the speed of the fan can be adjusted by swinging a cover wholly or partly over the solar cell.

No doubt the idea would work. However, Mr Dahly forgot two crucial points. The first is that even in 1967 hats were out of fashion, and have been ever since, so that the sales possibilities were minimal. The second is that, if you do insist on wearing a hat, when it is hot it is easier – and far cheaper – simply to take it off.

LESSON THREE: YOU CAN'T GET OWT FOR NOWT

We now come to that black hole of invention, the perpetual-motion (PM) machine. National Patent Offices used to accept perpetual-motion ideas because they did not wish to inhibit the free flow of invention. But there were so many on offer that the Offices began to demand working models, or attempted to get Parliament to define perpetual motion as frivolous.

But perpetual motion refuses to die. I cannot count the number of PM suggestions which I have been sent after presenting new technical ideas on television. I don't suppose anything I can say will change this, but I will try to spell out the problem once more. In future, if anyone sends me a PM proposal, I fear I will be unable to reply.

A perpetual-motion machine tries to get more energy out of a system than is put into it. This is not possible. A few seconds' thought will show that it is about as sensible as trying to get two litres of water out of a one-litre bottle. A difficulty is that many inventors acknowledge this fact, but simply do not recognize it when it applies to their own brainchild. Ignorance is not bliss.

Let me say loud and clear: perpetual motion does not exist.

Worse still: because any energy-transmission system involves losses of energy in friction and general inefficiency, even if PM did exist it could not be used effectively.

Let's look at some of the notions I have been sent.

- A ship cleaves its way through the water and displaces that water as it does so. Is this not a waste of energy? asks the earnest correspondent. If water were forced to run in a tube through the body of the ship itself, it could drive the turbines which power the ship and so keep the boat moving onwards for ever. It's a seductive thought – except that turbines work at only thirty per cent efficiency, so the ship would rapidly wallow to a halt.

- Or what about the suggestion that free energy is available after an office block is built? As the building rises, more and more potential energy is stored while walls and floors and ceilings are lifted hundreds of feet above the ground and placed in position. Why not tap this energy to light and power the building? A nice try. But there is no known way to draw off the energy. Even if you could do so, the only result would be to extract the energy which holds the building material up. Result: collapse of block.

- A slightly saner suggestion is that turbines could be installed in sewers to generate electricity from the water's flow. The water really is running downhill and there really is energy to be extracted from it. This is not a true perpetual-motion situation. Even so, if turbines take energy from the water and use it to make power, the loss of that energy will also stop the flow down the pipes, with horrific results.

There's no such thing as a free lunch. Forget perpetual motion.

Lesson Four: Big Names Don't Guarantee Success

Since the days of the South Sea Bubble, it has been standard practice to have a titled person, or a prominent figure, at the head of a company, whether the honoured name knows anything about the business or not. The implicit idea is that because he is such a respectable figure, he cannot be involved in anything at all speculative. This is nonsense, but even some of those who see through it from the start still fall for the ploy, or for the lure of vast riches promised by an invention.

Samuel Langhorne Clemens – better known as Mark Twain, the American humorist – was not fooled for a moment by the big-name game. He called the men who repeatedly appeared in it 'The Tone Inducing Committee'. But Clemens himself lost a fortune on a patent printing press. And he was the unsuccessful inventor of a detachable elastic backstrap for waistcoats and other semi-detached items of menswear.

There's a lovely story about Sir Isaac Newton which may or may not be true, but it illustrates the important psychological truth that a figure prominent in one area can be an innocent in others. Newton is one of a trio (the others are Darwin and Einstein) who contributed more to scientific thinking than any other human being. He discovered laws of optics, motion and gravitation.

The physicist had a cat, which he loved and for which he made a flap in the door of his house, so that it could leave and enter without hindrance. When the cat became pregnant and was delivered of a batch of kittens, Newton, one of the greatest intellects of all time, cut a second, smaller, hole in his door, so that the kittens could go in and out.

Lesson Five: Don't Waste Time

You've only a limited supply of time – they're not making it any more. While you fiddle around failing to pursue an idea, someone else may come up with the same notion.

Charles Darwin started to write a book spelling out his ideas about evolution. The subject was in the air at that time and friends repeatedly warned him that he must publish or be pre-empted. Fourteen years later, Darwin was still polishing the book! He was shattered to receive a letter from Alfred Russel Wallace, presenting the same theory. Darwin and Wallace agreed to publish their thoughts on the Origin of Species as a joint scientific paper and the following year Darwin came out with his book. The story has a happy ending, for posterity has always recognized Darwin as the true begetter of the theory of evolution.

There are any number of further cases where the same innovation has appeared simultaneously in different parts of the world, the best-known example being Thomas Edison (in the USA) and Joseph Swan (in the UK) developing the electric lamp. This is not coincidence. If there is a problem to be solved, a need to be met, it is inevitable that more than one person will be trying to find an answer.

Commercial rivalry has no room for the politeness of scientists like Darwin and Wallace. The competition is out to beat you. You may rise early, but they will have been up all night. Leave a promising thought on the shelf and the next time you dust it off and take it down, it may be only to throw it away because someone else has already patented it.

After you have learned from other people's inventions what you can about the causes of success and failure, you must think about your own ideas. Do the lessons apply to you? A major cause of wasted time and effort comes from not finding out, before you become fully committed to it, that your idea has already been thought of. It may not be on sale, and you may never have seen or heard of anything like it, but that is not enough. You need to know its patent position. Is it new or not? As soon as you have worked out in principle what you intend to invent, you must act on the next Rule.

Rule Four: Check for originality

If the idea behind your invention has been patented at any time in the past, even if that patent is long dead, you cannot be granted a fresh one. This makes observation of Rule Four absolutely essential. Check *before* you start developing your original concept: it could save you years of wasted time and a lot of research money.

At this concept stage, you probably aren't ready to spend much cash on the idea. Such as paying for the services of trained examiners at the Patent Office to check if the product has already been invented. So what are you to do?

The answer is to visit the British Library's Science Reference and Information Service (SRIS) in London's Southampton Row. Housed in a magnificent Victorian pile, this reference library is handsomely vaulted and skylit, and lined with two tiers of cast-iron balconied galleries. Here, in the Victorian tradition of making knowledge freely available to all – a tradition which a mean-minded age is throwing away to its own loss – an initial manual search can be carried out for nothing.

The reading rooms are open to everyone without appointment, and all you need to do is sign the entry book to confirm you will abide by SRIS rules. About forty per cent of the callers are academics, patent agents or consultants and the rest are a mixed bag, including freelance inventors. The library is a well-used place – it has 150,000 visitors a year – and rightly so.

If travelling to London is inconvenient, there are eight provincial patent libraries which also have UK and European patents available for inspection (details are on pages 224–5). The collections aren't as comprehensive as those of the SRIS in London, but they'll get your research off to a good start.

I won't pretend that making a search is easy. You have to find a way of telling the librarians which area you are interested in without letting them know exactly what you are up to (technically, that would be disclosure of the idea, but the librarians won't try to trip you up). The classification system covers such a vast area that it can be confusing but, again, provided you don't assume that the librarians will do the search for you – the SRIS expects payment if you go that far – they are very helpful.

The initial subject search is made by looking in the catchword index. Say that the idea is for an independent electricity supply, you will find yourself gradually moving from details of stand-by generators to solid-state devices and inverters. Eventually you end up in the stacks themselves, pulling down volumes, inspecting individual past patents to see what already exists; or maybe using the on-line public-access computerized material or microfiche catalogues.

If you do discover that you've been beaten to it, it's distinctly annoying. On the other hand, you can congratulate yourself on having found out now, rather than later, after months of effort and an uncomfortable amount of expenditure. Of course, if you are lucky, you will find nothing.

At this stage, ask if you want to employ a Patent Agent or the SRIS or Patent Office to make full inquiries on your behalf. There's no golden rule. If the invention is likely to be of small commercial value, it's probably not worth a paid search until it has been fully developed. If the idea shows signs of being a potential money-spinner, however, an early professional search is worth considering. Money-spinners have a nasty habit of being money-eaters in the development stage. Perhaps you shouldn't launch into development solely on the basis of an amateur search at the SRIS.

Whichever course you follow, don't forget to keep quiet about the idea until you have protected it. Don't show it to other people, don't talk about it and above all, don't publicize it. Otherwise you may find the patent application denied, on the grounds of previous disclosure.

CHAPTER SIX

Improve the Idea

Many inventors react to the mildest criticism of their latest products as if they were the mothers of new-born babies. They slave away, often for much longer than the nine months' work required for a baby. They know that their non-inventing friends have not the slightest idea of what it takes to deliver their brainchild, and any well-meant remarks that the offspring resembles an unliked grandparent (or earlier invention) go down like a lead balloon.

This is understandable – but misguided. Belief in the idea is essential, but belief that you are infallible is silly. No idea exists which cannot be made better. If this were not so, the whole business of invention would end.

Rule Five: Build a working model

When you have checked for originality, the next step is to test if your invention works in practice. You can do this by making a working model. Let's say you have dreamed up a long-arm power device which lets tradesmen and DIY enthusiasts paint ceilings and remote objects without standing on ladders or trestles. There is a motor with various electrical leads running to the rotary working head of PaintArm, as you have decided to call it. All is going well.

It is only when you put the working model together that you realize you have a major problem routing the wiring. It is quite astonishing how different the 3-D version of an invention can be from its 2-D representation on paper. This is where a working model comes in useful, as it performs the following functions.

- A model shows what parts are needed in manufacture. In this case you solve the wiring problem by putting the wires in a channel running up the centre of the handle, out of harm's way. To eliminate the danger of insulation-fray where the loom of wires enters PaintArm's flexible operating head, you pass it through a padded universal joint.

But will you place the rest of the wiring in a channel drilled down the centre of the wooden handle? You discover that drilling wood is technically not feasible, so you choose to make the handle of plastic. Will this handle have its central hole moulded in during manufacture? Or will it be made in separate longitudinal sections, grooved down their length, with the two halves being glued together after wires have been laid in the channel?

• A model helps to work out costings. Just one of your decisions, to go for plastic handles, determines one set of costs. A second choice, to have the channel moulded, fixes a second. The universal joint did not appear in your original concept. How much will it cost, and will it complicate the assembly procedure?

In a single product, there may be two or three dozen constructional problems to solve and price. You cannot afford to be ignorant of the answers.

• A model gives you the users' viewpoint. You heft the redesigned working model in your hand and find that its head is cumbersome in use. Fortunately, this difficulty can be resolved by putting the motor in a different position on the long-arm handle so that head and motor are nicely counterbalanced.

You would not have known this problem existed unless you built a working model.

• A working model clarifies the idea. When you talk to a patent agent, a real-life version of PaintArm gives him a clearer notion of what you claim as unique and unprecedented. As a result, your patent application is likely to be improved in accuracy, as well as being cheaper to draft.

Now you can advance to the prototype stage, itself based on the working model. The prototype is the first real version of the invention. With luck, it will be quite close to the final product, shiny and full of promise. Make it as professionally as you know how.

You will eventually find yourself dealing with manufacturers who need to know that the design is good and its costings are soundly based. They won't be interested in spending their own money on improving it. A good prototype answers most of their questions and (leaping ahead) a good pre-production model answers all of them.

Pre-production means that you have organized full manufacture of a number of batches of the invention. This gives you working specimens which can be used for sales purposes, market testing and advertising. It also means manufacturers can be assured that you have ironed out all the production problems.

When it comes to distributors, who may have no expertise in your product area, you have to convince them that your idea will sell. They will find it far easier to understand a prototype or pre-production model than a drawing. One demonstration is worth a thousand words.

I offer this encouragement to produce a working version of your idea in the full knowledge that there can be difficulties in turning it out. In response to my survey. Inventor B says: *'It seems a permanent situation in the world of precision engineers that those who have the ability to make up a model, with the accuracy needed to demonstrate the state-of-the-art advance which underlies the invention, do not have the time to devote to it. And those who have the time do not have the skill to produce an attractive and convincing model!'*

Few organizations will make an inexpensive prototype for you. Some of them advertise in the magazine of the Institute of Patentees and Inventors. Some university departments will also make prototypes, especially at technical universities like Brunel.

People do have the skills, especially those who have been cast aside by industry, but in many cases they lack the right equipment. Even I haven't got the facilities to produce prototypes on a grand scale, though mine is a fairly well-equipped workshop.

I can carry out small-scale engineering. For instance, I built a diagnostic device which measured how bodily weight is distributed when you are standing up. It was rather like the instrument Samantha Haines produced for stroke victims (see page 38), only more elaborate. From the pressure of the patient's feet, you could diagnose the amount of hip-joint displacement, for example, and calculate the remedial treatment he or she needed. However, anything that requires fairly heavy engineering is quite beyond my capability, and unless you are equipped for this kind of work, you will have to pay through the nose for it.

Inventor G states: *'It is difficult to obtain small quantities of materials for prototype work at reasonable cost.'*

Again, it's a question of money. You *can't* get small quantities at a reasonable cost in the UK. Because of computerization, and the malign influence of accountants, the engineering industry now supplies only standard amounts and quantities, none of which are what you are likely to require. When I want materials, I go to European manufacturers or distributors, who are quite happy to sell me small quantities. If you need a contact in this area, try the commercial attaché at the embassy of the country concerned.

A newer source is the countries of the former Eastern Bloc. They have an old-fashioned engineering structure which still supplies individual orders and they are desperate for business. But I can't promise speedy delivery!

Inventor B has an honest confession to make: *'I failed to take enough control over prototypes and demonstrations being undertaken by companies trying to promote the idea e.g. fifty machines were equipped with prototypes modified by hand, some of which worked well, others not at all. This obviously created major problems when out on test.'*

The writer mentions 'prototypes modified by hand'. When you use hand-modified equipment, you are bound to get failures. Hand-modification is not as good as machine work, for it is often inaccurate. This problem makes quality control absolutely essential. There are never many items involved at the prototype stage (fifty, as in this case, is a high number), so quality is easily achieved. Every item must be checked for durability and workability *before* any outsider sees it.

Case history

Earlier I briefly mentioned Stephen Harding's ten-speed hub gear for bicycles which is worked by air bulbs on the handlebars. I have tested this invention myself. It is a real advance on derailleur gears, with their awkwardness and tendency to jump the cogs.

Stephen cut the internal gears of his prototype from steel – a material conveniently available to him – fully intending that production models would be made from plastic. His invention was turned down unseen and untested, on the grounds that the public would not want it. I admit to speculating here, but it seems to me that – albeit for the best of reasons – Stephen made a mistake. The prototype looked somewhat bulky and heavy, and that was enough to frighten off an industrial company which was only half-interested anyway. Although it doesn't matter so much in the working model, it's important to use the right materials in the prototype.

Let's return to making a working model. This doesn't sound

Gerry Lloyd's improved supermarket trolley. His introduction of a trailing fifth wheel stops trolleys and airport baggage carts drifting off course. (*Picture: Gerry Lloyd*)

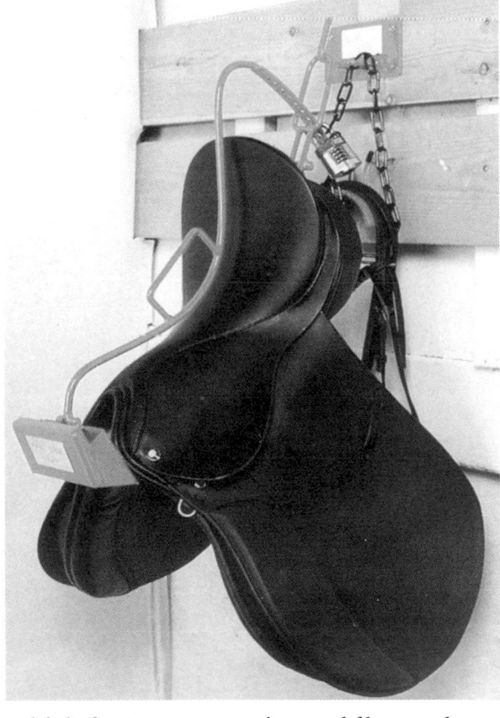

Vivi Hagrup with her Portack-et-Lock which fastens expensive saddles and bridles – tack – to the walls of stables and horse boxes, securing them against theft. (*Pictures: Vivi Hagrup*)

Innovation pays off. On the left, the building where Frank Arthur, inventor of Envopak, started out selling his idea: on the right, Envopak's impressive premises today. (*Pictures: Envopak Group Ltd*)

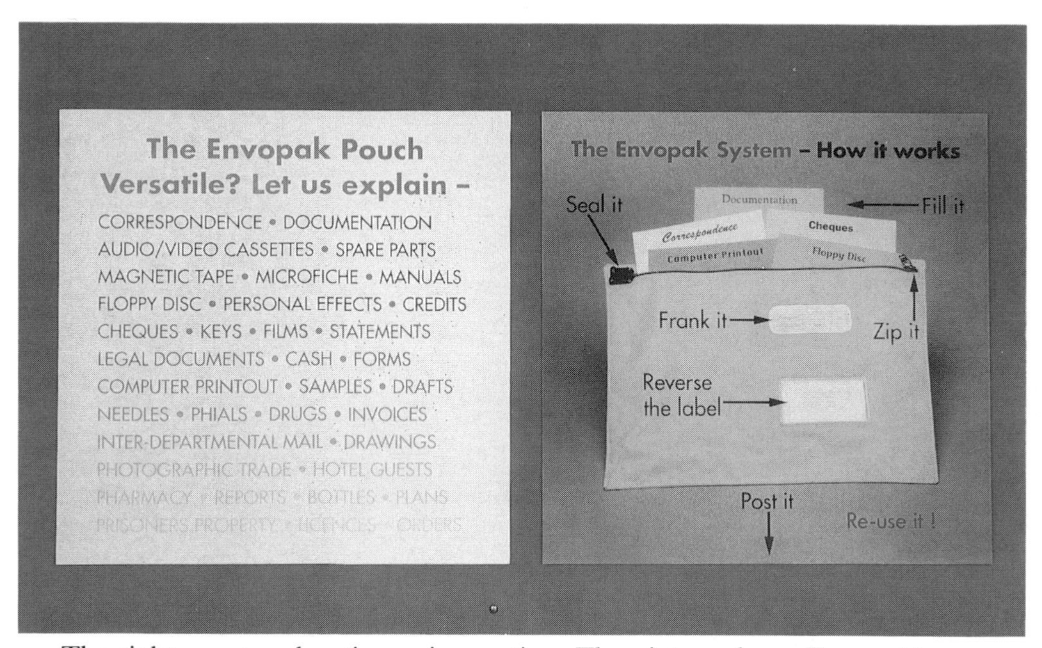

The right way to advertise an innovation. The picture shows Envopak's envelope and how it works; the words explain where it can be used to advantage. (*Picture: Envopak Group Ltd*)

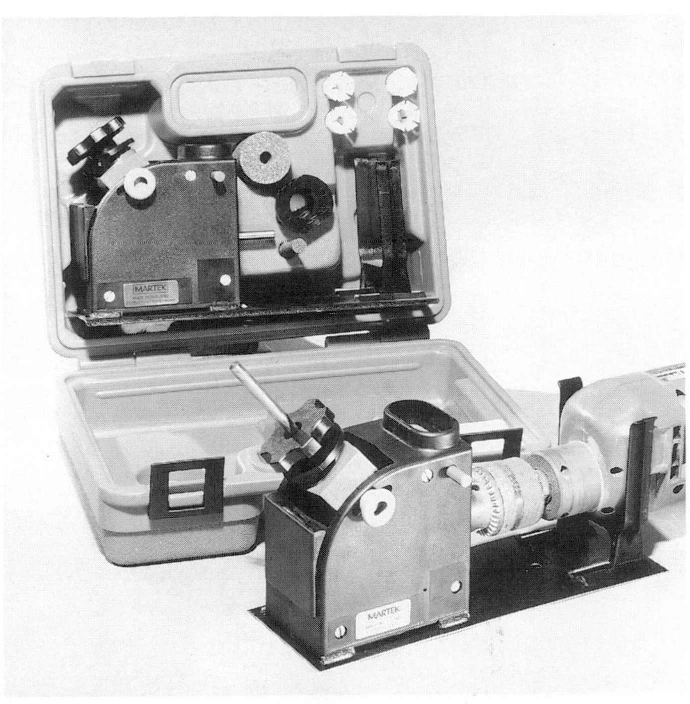

Meet a need. Tom Ketteringham's Professional Drill-Bit sharpener saves money for DIY enthusiasts *and* improves their results. (*Picture: Martek Ltd*)

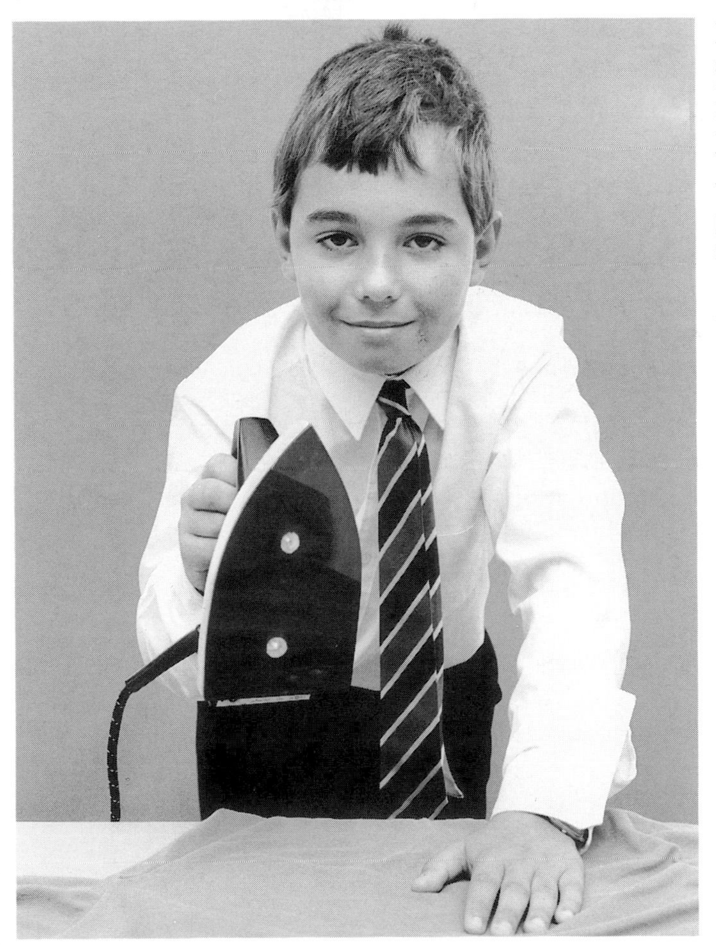

Innovators can be any age. Twelve year old Ross Carden from Monkwearmouth School with his safety iron which switches off to prevent it accidentally starting a fire. (*Picture: Young Engineers for Britain*)

Changes in the school curriculum are producing more female inventors. Sixth-former Samantha Haines from Cheltenham (boys!) College with her physiotherapy aid for stroke patients. (*Picture: Alan Miers*)

The Maltron typewriter keyboard. It is ideal for computers, quick and comfortable to use, but faces the problem that potential customers have already trained on the old QWERTY layout and are hesitant to change. (*Picture: Alan Glenwright, Take A Break*)

Where to check the originality of your invention. The magnificently galleried Victorian interior of the Science Reference and Information Service Library in London's Holborn. (*Picture: The British Library/Science Reference and Information Service*)

Where European patents are granted. The sleek building of the Patent Office in Munich which handles applications from seventeen member countries. (*Picture: Petra Flath*)

The invention that got away – a cautionary tale (page 78). Bob Symes with his diesel-hydraulic model locomotive. (*Picture: Associated Press*)

Towards the rear of the locomotive's power unit is the control valve that Bob Symes failed to patent. It is now in world-wide use on automatic lathes. (*Picture: Bob Symes*)

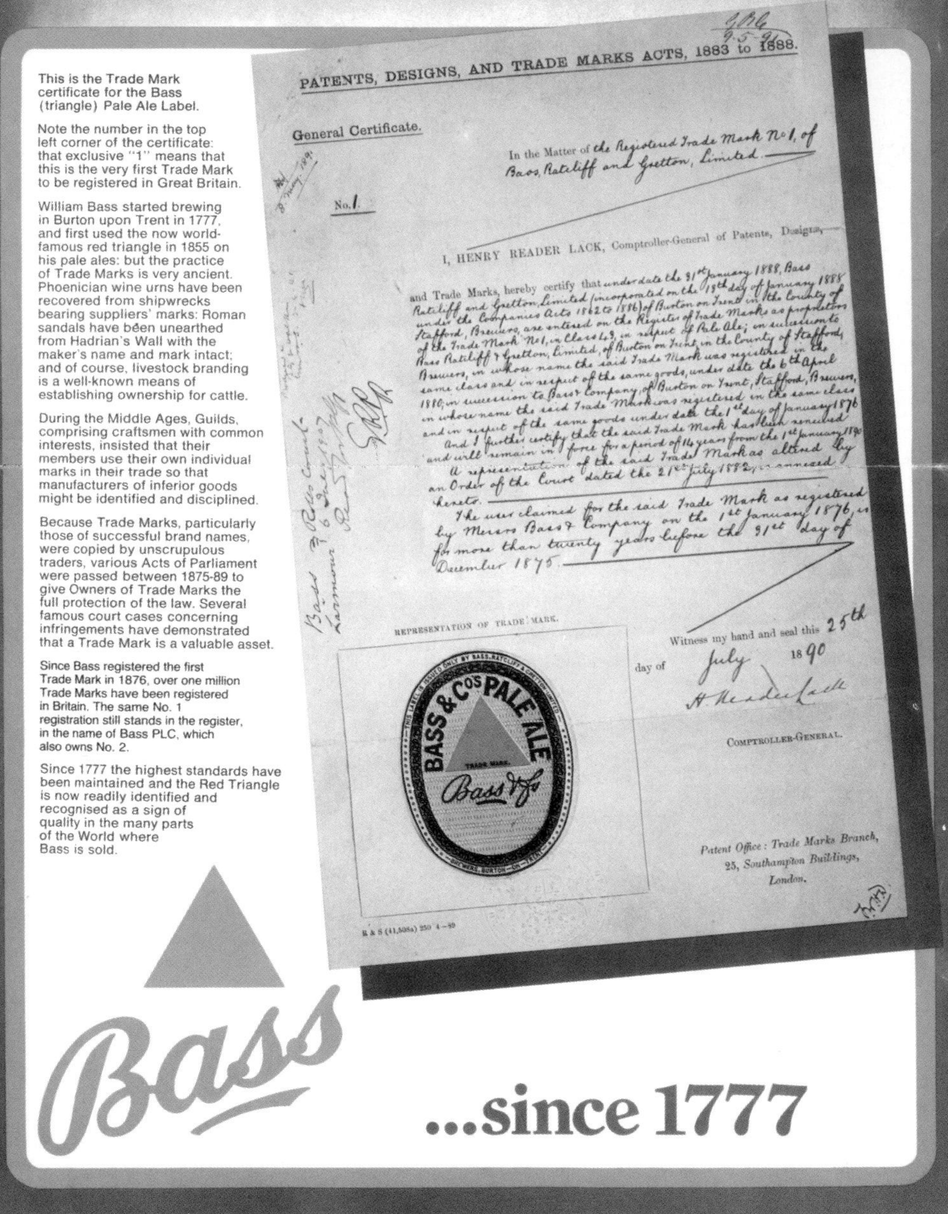

A fascinating poster produced by the Bass brewing company, showing (near the centre of frame) Britain's oldest registered Trade Mark, still in use today and highly valued by the company. In colour, the triangle is printed red and the pattern behind it is also red. (*Picture: Bass PLC*)

difficult until you remember that even the most mechanically minded innovator is not likely to own a vacuum forming-machine, or have access to different formulations of plastic materials. Profile cutters are in short supply in the home, as are electric furnaces. Fortunately, there are places where these problems can be solved.

Innovation Centres

Some Innovation Centres are heavily business-oriented, specializing in help with premises, marketing and legal questions; others concentrate on the practical problems of invention itself, such as prototypes and patent searches. Set up to help and support innovators living in their catchment areas, they are financed by a mixture of sources – government, local authorities, chambers of commerce, and the European Union.

The Centres began life rather slowly at the end of the 1970s, only a handful being set up in the first five years. Now there are nearly thirty spread across Britain, although there are none north of Loch Lomond or west of Cardiff. (See pages 231–3 for addresses.)

What can you gain from an Innovation Centre? Initially: practical help in turning an original idea into a working model and a prototype. The key figure is a person variously described as counsellor, advisor, consultant, director or manager. He (or, in a very few cases, she) acts as a guide, philosopher and friend to a diverse clutch of innovators; a guru, no less. If you don't get on with the counsellor personally, you are not likely to make good progress. But he is usually the most affable of people. He can evaluate ideas, give guidance about patenting and other forms of protection, help with arranging licensing, and point you in the direction of the best specialist assistance (which may even be supplied by the Centre).

A word of caution: Innovation Centres are not magic. They can't turn a weak idea into a strong one. Nor can they guarantee that even a good and necessary invention will be picked up by a manufacturer, or will sell if you choose to manufacture it yourself. But they can certainly tilt the odds a little in your favour.

Perhaps you dislike the idea of an Innovation Centre. You would rather keep everything under your own control, and not let outsiders see your progress with your working model. There are ways to do this, but they have the following disadvantages:

- The model will be built at your own expense. You are not only throwing up the chance of subsidized help from one of the

Centres, you are committing yourself to the hire or purchase of the equipment needed to make the model.

• You also have to hire expertise if you do not know how to operate some of the machinery or lack certain manual skills. This can be expensive.

• If you bring in paid experts, you may be in greater danger of losing confidentiality than you would in an Innovation Centre. There's a freemasonry of shared struggle in the Centres which does not exist elsewhere; nobody in them is likely to steal your idea. It is good policy to let employed outsiders see only part of what you are doing. An even better approach is to pick up private experts as you go, choosing them carefully of course.

Learn About People

Part of the inventor's art that is hideously undervalued, indeed is never discussed, is understanding other people – empathy. Inventors need to be able to persuade, to cajole, to influence, to get what they want from people who don't necessarily want to give it to them. Many are so carried away by the beauty of their idea that they cannot see how it looks to outsiders, or how *they* look to outsiders.

The result, too often, is an inventor left sobbing in the corner, licking wounds which are largely self-inflicted. 'I trusted them and they let me down,' is the cry. Yes, but did you take any time to study the person you were dealing with? 'They wanted to know too much before they put up any money and when I wouldn't tell them, they walked out on me.' Yes, but if you had already protected your invention with a patent (see Chapter Seven) were you being unduly suspicious about an honest commercial query?

You have to find the right balance between being over-confident and being continually suspicious. It takes practice, but when you have the knack, you'll be astonished at how much help you can get from others, how many people you can find with the knowledge to improve parts of the invention where your own skills are limited.

Finding such assistance has become easier in recent years, for the sad reason that thousands of skilled people in their fifties have been made redundant – 'taken early retirement' as the cant phrase has it – and many of them are delighted to work on a private project. Using their skills reassures them that they are not, after all, worthless as individuals. Ask around locally – you'll find the electronics wizard who can improve your circuitry or design it from scratch, the plastics expert who can turn out your mouldings, the versatile telephone fixer and many other skilled individuals.

Collaborate

Now the situation becomes trickier, because I am going to tell you to give away part of the value of your idea when you have been taking extreme measures to protect its confidentiality. It may be a matter of assigning only five per cent of the project value to your collaborators but, if you don't do it, you are putting a heavy burden on human nature. Your skilled aides may be completely honest, but how long can they keep quiet if you have something red-hot under development? A financial stake in what you are doing is the most persuasive way of buying their silence. Don't let them tell even their partners what you are up to.

It's also important not to haggle over who contributes how many washers or writes how many letters. From the start, establish a friendly but formal understanding in writing. If you can't agree on this, you shouldn't be working together.

Improve The Design

The time has come to improve your idea in a different way. Stand back and look at the appearance of the working model. It may operate beautifully but look like something the cat brought in. Who will want to buy it? Good design is as important as function. Ideally, appearance and use go hand in hand. But in practice, an invention's elegant casing too often covers a disorganized interior. Or because of the way it is designed, a superb product cannot easily be used by the consumer.

Examples include office buildings which look beautiful on the outside but turn out to be impossible to work in because they are too hot in summer and too cold in winter, or the roof leaks. Again, chairs with great visual appeal sometimes cause backache, while books or newspapers using elegant type styles can be difficult to read and are therefore ignored.

Case history

This book is printed in what is called a 'serif' typeface; each of its letters has tiny flourishes at its extremities. The capital 'A' stands on tiny feet; the small 's' has points at the end of each curve.

The design-for-its-own-sake brigade goes instead for a type style called 'sans serif'. These paragraphs are printed in a 'sans' typeface. As you can see, it has none of those apparently useless extra bits of print. It looks clean and unfussy. But tests on how easily we take in a page of print show that serif type is more comfortable for the human eye to handle in quantity. This is because our brains work most efficiently when we have redundancy – more information than we need. Redundancy lets us pick up the clues we find most helpful.

So sans serif is fine for short passages of print, where it makes a dramatic impact. On the other hand, read an entire American newspaper printed in sans, and you rapidly develop temporary brain damage.

Another graphic design trick is to print, say, thin blue lettering on a black background, or green lettering across a black and white photograph. The result looks good at a distance – but just try reading it!

Similarly, some manufacturers fail to understand that design is for people to use. Do not be one of them. The video recorder is a classic example of a product which desperately needs a new look. They might be a wonderful invention, but video recorders drive most of us crazy. VCRs are the principal piece of household equipment which we cannot work properly, and research shows that eighty per cent of owners have difficulty programming their machines.

There are several design problems common to most VCRs:

• The remote control for my Sony machine has no fewer than sixty-four buttons plus a rotary 'swing-shuttle' wheel at the base. Airline pilots have less complicated control panels. Sony itself is conscious of the problem. It puts a special sticker on the back of the remote control, which gives step-by-step instructions on how to work the timer!

• The tiny size of the buttons makes it difficult for anyone elderly, anyone with poor sight, anyone with shaky hands, or anyone a bit baffled by the machine, to use those controls accurately or effectively.

• Some video recorders measure the progress of tape through their innards by using an inscrutable system which makes it impossible to find a desired point on the tape. Why can't they work in minutes and seconds?

• The machine is intended to sit on a shelf underneath the TV set. This makes it difficult to get at: you have to go down on hands and knees to see the controls. Why shouldn't the recorder, which needs programming every time it is used, be on top of the TV where you can reach it standing up?

• The controls are built into the face of the recorder. As it sits low down, near the floor, these are extremely difficult to work. Why aren't they on the top surface of the machine, so that you can push the buttons from above?

I suspect that video recorders might move off dealers' shelves faster if ordinary citizens could work them comfortably. Even if this were not true, professional pride ought to make manufacturers come up with a more practical design.

Finally, it seems evident to me that one answer to the whole problem would be voice-operated VCRs. There is no reason why existing technology cannot make equipment which accepts verbal commands, eliminating all the confusing button-pressing.

Case history

I was ready to advance a detailed explanation of how voice operation could work when I discovered that Philips was launching the Voice Commander. It responds to spoken commands and can 'control virtually any TV, VCR and satellite tuner'. You speak into a microphone, giving the day, channel and programme times. The recorder recognizes four voices (one for each member of the statistically average family) and can even be told not to accept adverts.

It has taken twenty years to get to this point. But why did the problem arise in the first place? The answer is, because the difficulties we suffer as consumers were not apparent to the companies' technically minded designers. With years of experience behind them, they already knew how VCRs work. They couldn't see that there was a problem.

The lesson for you as an inventor hoping to sell your product is to have it assessed by people who represent the eventual user. The computer software business calls the process 'beta-testing'. Trial disks of new programs are sent to consumers. They report on 'bugs' in the programs, and the companies then try to eliminate them before the final products hit the market. It doesn't matter whether you are inventing a software programme or a new lavatory brush. Get away from the inventor – in this case, yourself – and find out the reactions of the people you hope to sell to.

Use your time with working models and prototypes to sort

out problems which may arise when you attempt to place the product with manufacturers and distributors. The moment one of them says, 'It's a good idea BUT . . .' it's too late for you to come up with extenuating pleas about the next stage of development.

Rule Six: Don't attack established interests

The early stage with working models is also a time to sit back and contemplate the totality of what you are doing. You have the invention. It works nicely. But has it a place in the real world? Out there – on the alien planet where ferocious commercial predators stalk the landscape, eager to devour any morsel that falls to their sharpened claws – what are your chances of survival?

Could it be that you have a breakthrough with your idea, but it can only succeed if it destroys a range of existing products? If so, you have a serious problem. Is it likely, for example, that the makers of a range of cars will welcome your suggestion for an entirely new engine, wonderful though it is? Will manufacturers of biodegradable plastic bags want an innovation which destroys their market? How many consumers who have mastered one computer programme will be willing to invest in and learn a different one – even though it is faster, more versatile and easier?

Case history

The standard QWERTY typewriter keyboard was not designed with efficient use in mind – the very opposite. The keys were arranged in that order in 1872 to slow down skilled typists who were causing typebar jams by hitting the keys too rapidly.

Stephen Hobday, a successful inventor and electronics engineer, joined forces with Lilian Malt, a keyboard training specialist, to product the alternative Maltron keyboard. Its keys are in four blocks: two larger ones for the fingers of each hand and two smaller ones for the thumbs. The blocks are curved in two directions to match the ways fingers stretch and reach. The keyboard bears a passing resemblance to an inverted turtle shell.

Lilian Malt developed a letter arrangement which is different from QWERTY. Frequently used letters sit immediately under the fingers (see photo section). The new layout is easier to use and typing speeds increase. Typical comments are: 'It took me only three days to learn the key configuration. In those three days my speed and accuracy went up dramatically' and 'I find it so much easier than my old keyboard.'

The first development work was carried out on the Maltron in 1976, when conventional typewriters still ruled the office. But the established interest that it needed to defeat was not the manufacturing firms which dominated the typewriter industry. The problem was the fixed attitude of hundreds of thousands of typists. They had learned the existing QWERTY method, many of them becoming highly skilled at using it, and they weren't about to throw all that effort away, however good the Maltron letter arrangement might be.

Maltron didn't have the fighting weight to overcome this inertia. Recognizing this, Stephen Hobday eventually installed a switching system which allowed operators to use the old QWERTY arrangement with the new design of board if they wished.

The product was good, extensively beta-tested, and its price was reasonable. But sales stayed at a low level. They were enough to justify carrying on, but not large enough to make mass production worthwhile. In short, buyers' resistance did not break down.

But then the situation changed. Word processors and work-stations suddenly became much more than a gleam in IBM's eye, and the manual typewriter vanished from offices. There is now a generation at work that has never seen a typewriter – and who, ten years ago, would have believed that?

The logical Maltron keyboard is well suited to use with computers, and the QWERTY key arrangement has no logic left in it when there are no metal keys to clash together. Sales are increasing and Stephen Hobday has his son and nephew helping to turn out keyboards because he cannot cope with demand on his own.

Repetitive Strain Injury (RSI), which the Maltron was originally designed to alleviate, has become front-page news. Using the Maltron, even with the QWERTY layout, can keep people with RSI at work instead of condemning them to disability. Indeed the manufacturers have letters from buyers whose careers have been saved by using the new keyboard.

Stephen Hobday and Lilian Malt's invention is succeeding because it has *met a need*, although it still challenges some established interests.

• Get a fresh view. Makers of multimillion-dollar feature films try them out at private viewings before showing their efforts to the public. Giant companies call in firms of independent consultants to advise them how to save money and improve production. Onlookers see most of the game.

I've suggested that you should make a conscious effort to stand back from your invention and enlist outside help. Here comes the same advice again, but with a twist to it. Once you have your invention legally protected – and only then (see Chapter Seven) – ask one or two technically minded people you trust to see what improvements they can make to it.

I've been privileged to be consulted as an *ad hoc* adviser on many occasions. I can't claim that anything I have suggested has catapulted the inventor to fame and the fleshpots, but I hope I have helped to make a few things more saleable.

Case histories

• Fire escape. The fire escape looks like an aluminium drainpipe fixed to the outer wall of a house. When you pull a pin, the two halves of the 'pipe' swing apart to become vertical struts, and folding arms inside it straighten out to form the rungs. It's a neat idea to help people escape fires by climbing out of the window on to the ladder.

The original invention was clamped to the wall with special expanding bolts. This could be dangerous because expanding bolts left unattended (as they would be) can become loose. There's no guarantee they will stay in place. I suggested that a metal plate should be added on the *inside* of the wall, with the exterior bolts extending through the brickwork to fix into it. So, when someone fleeing a fire throws all their weight on the top end of the escape, it is safely anchored.

The release pin was originally outside the window. The window had to be opened to operate it, which created a period of time during which a fire raging inside the house would be fed with fresh draughts of oxygen. It seemed sensible to put the release inside that that you open the window only when the ladder is fully in position.

• Gun sight. This is a simple device to aid clay-pigeon shooters. The problem for poor marksmen is that the clay flies so fast that, unless the gun traverses slightly ahead of it, the shot misses its target. The new ring sight is 'angled off' in such a way that when you see the target through it, the gun is automatically in the right position to knock it down. The sight is adjustable for the range at which you are firing. Dashed unsporting, what? But very effective.

I suggested that the inventor enlarge the size of the ring. He had designed a small one which was quite difficult to sight through and which gave the precision required for shooting with a rifle. A larger ring was less accurate, but much easier to use; when you use a shotgun, which fires spreading pellets, rifle accuracy is not required.

• Brake light. It's a source of continual surprise to me that cyclists are happy to ride around at night almost completely unable to signal to other road users what they are going to do next.

The light tackles one of the problems, signalling that the cyclist is slowing down – just as a car's brake lights do. A mercury switch inside the rear lamp activates a light bulb as the bike comes to a halt. But there was a minor flaw in the design: because deceleration is uneven, the bulb flickered instead of shining steadily. All it needed was a delay circuit next to the mercury switch; this evened out the flicker to produce a constant glow.

• Ready-Stitch. We've already come across surgeon Riad Roomi's idea for eliminating stitching with needle and thread from the treatment of wounds. It's almost impertinent of me to claim that I made a contribution to the invention, but I hope I was able to add a couple of thoughts. The first was truly trivial. When Mr Roomi wanted to demonstrate Ready-Stitch in public, he had a presentation problem. Real human wounds could hardly be used, except on video or film, which was not always appropriate to the occasion. I proposed that he show an orange with Ready-Stitch plastered over a cut on its surface.

Secondly, I knew that surgeons sometimes draw a grid pattern on the patient's skin before making their incision; it helps them to line up the two sides of the wound accurately as they stitch it together again. I suggested that Ready-Stitch should include a built-in graticule – a grid pattern printed on clear plastic – to produce the same accuracy.

There is nothing miraculous in any of this, is there? It is simple for any outsider with a degree of knowledge to help an inventor. Outsiders see weaknesses in design, or approach, more easily than someone who has been keeping their nose to the grindstone for months or years.

Use the skill of outsiders, as long as you don't compromise the security of your idea. And remember to obtain patent protection. The cases described here were already at the 'Patent Applied For' or 'Informal Application' stage or further on, when I first publicized them.

CHAPTER SEVEN

Patents – the Theory

What is a patent? A privilege granted by the State, it is a monopoly giving you the right for a maximum of twenty years to stop others using your invention without your permission. Let's begin our look at patents with the message that runs through the next three chapters:

Rule Seven: Learn the patent system

If you hope to be awarded a patent, you must keep your invention secret until your application is filed. You mustn't write about it, or make it known by showing or using it in public. Don't let yourself be interviewed about it by the local paper; don't demonstrate a model. All these actions constitute 'disclosure'. This means that a patent cannot be granted and, if by chance you should succeed in getting one, its validity can be challenged. In the words of the war-time poster: 'Be like dad. Keep mum.'

There's one exception to the rule. If you tell someone about your invention in confidence (and that means getting a promise of confidence *before* you reveal all), you can still obtain a patent. It's OK to talk to patent agents. You don't have to be totally paranoid about secrecy.

Problems With Patents

A patent is valuable because it proves to manufacturers and buyers that you own an innovation. Like a horse, it can be bought, sold, hired or licensed. You can use it yourself, or allow other people to use it by agreement; you can sue people who infringe it.

It is not as strong a defence as many people believe. Neither the Patent Office nor the State enforces it. In fact, the first refuses to take sides in any patent dispute – reprehensibly so, in my opinion. If the State has made a contract with you, for which it draws increasing dues all through the patent's life, should it not assume some responsibility for the integrity of that contract?

The sad fact is, you have to police your patent yourself. That includes attacking infringements of it in the courts. Courts are time-wasting, soul-searing places and your opponent may have £1000 in his fighting fund for every pound that you can lay your hands on. You'll have noticed from reports of cases in the civil courts, it's the side employing the most expensive lawyers that tends to win.

The best strategy after you have gained the patent is to get into the market-place with a good product and hold your own by a process of continual improvement. That's what wins the day. Try to stay clear of the law.

There are limitations to patents. You cannot stop other people taking out a further patent if they discover an improvement to your idea, even though their superior flytrap destroys the value of your own version.

The State – those friendly people who mis-run the country but in this case are right – has a purpose in handing out patents: to encourage industry and make the nation and all of us richer. It wants both flytraps to slug it out toe to toe in the market-place, and may the better trap win. The State's long-term thinking is that the prize of a temporary monopoly will persuade inventors to release details of their innovations for the common good instead of keeping them secret.

Three generations of a family of French doctors held close to its collective abdomen a new design for obstetric forceps which made childbirth safer and easier. It is not known how many babies died because knowledge of the forceps was not released to the medical profession, but there is no doubt that the family's secrecy was bad for progress.

The Origin Of Patents

The word 'patent' comes from the Latin '*litterae patentes*', which means 'open letters'. Letters patent were documents issued by the Crown with the Great Seal attached at the bottom by ribbon. The documents were 'open' because they could be read without breaking the Seal. From the thirteenth century onwards, letters patent were used to hand out government instructions of all kinds (they're still used to appoint judges and members of the House of Lords). Gradually their use widened to include promoting trade or raising money for the Crown.

Other countries had the same approach. The Italian architect Filippo Brunelleschi, renowned for designing the wonderful dome of the Cathedral at Florence, invented lifting gear for the barges he used to carry marble to building sites. In 1421 Florence gave him a three-year monopoly on the idea.

Italians were quick off the mark about patents because they were the first mercantile civilization in Renaissance Europe. Patents in those days, as now, were about making *money*. Italians led Europe in banking, and invented pawnbroking. Their first centre in London was called Lombard Street after the north Italian moneymen who set up their exchanges there. 'All Lombard Street to a china orange', as an expression of betting certainty, still enshrines their success.

The English learnt a rapid lesson from this early trading version of the European Union. In the reign of Henry VI, the first English patent for an invention was given to John of Utynam, to make glass for church windows. John supplied glass to Eton College, no doubt for its famous chapel. He was required to train other people in his process, an early version of the idea that patent knowledge should be spread.

By the nineteenth century so many patents were being awarded for inventions that the word 'patent', on its own, came to have its present meaning.

Alas! Progress has abolished the romance of the patent, engrossed on a great sheet of parchment with the waxen Seal dependent at its foot. Nor is it awarded any longer by the sovereign. Owners of patents are properly not even patentees any more (though most of us continue to call them so) but grantees, because the award comes as a certificate of grant.

What Is Patentable?

One of patents' curious aspects is that, for centuries, there was no real definition of what they actually are. Earlier British laws said they were 'any manner of new manufacture the subject of letters patent', which is as near a useless circular definition as you could hope to find. It meant that the courts made the final decision. Law lords based their judgements on precedent: what other courts had already decided. This got no one nearer to a practical definition of what is patentable. In a typically British way the system worked well enough, but without making much sense.

In 1977 the United Kingdom ratified the European Patent Convention (EPC). At last, this spelled out the minimum requirement for a patent. To be patentable, an invention must be new and capable of industrial application. It must also not be obvious to anyone of normal ability and skill who knows the technology involved. For instance, chemists continually develop compounds in their search for effective new materials in the field of agriculture. Making the chemicals themselves is almost a number-crunching exercise, so routine is it. Any competent chemist can do it.

This is why the inventive leap forward with a new agrochemical is generally regarded, not as its formulation, but as the discovery of its value as a pesticide or fertilizer.

What Can't You Patent?

Several classes of item are ruled out:

- Discoveries and scientific theories. A scientist from Iowa State University proposed blowing up the moon to stabilize the earth's climate. Removing the moon from the sky would get rid of tides (which have climatic effects) and slow down the earth's long-term wobble as it spins through space. Stopping the wobble which creates a phenomenon known as the Milankovitch effect could prevent the earth moving into a new Ice Age.

  Think of the problems which exploding the moon might create. No patent on this idea can be granted in Europe. Or anywhere else, one hopes.

- Works of artistic creation. These have copyright protection instead. Computer programs come under the same heading, though current law cases in the USA make it possible that in that country the 'look and feel' of a program may become patentable.

- Methods of treating the human body. These include surgery or therapy or methods of diagnosis – they are not regarded as inventions with an industrial application – but exclude products designed for use in any of these methods. An invention such as Ready-Stitch escapes the ban because it is a product, not a treatment.

  In an allied biological area, patents will not be granted for new animal or plant varieties (although certain plant varieties may be protected under a scheme administered by the Plant Varieties Rights Office).

- Inventions which are against the public good. Anything that might be expected to encourage offensive, antisocial or immoral activities will not gain a patent. Fifty years ago there was a consensus about what such evil activities might be. But as western civilization stumbles into its third millennium, definitions seem rather more debatable. What is immoral to you may seem to someone else the exercise of their inalienable rights as a free citizen. If you want your immoral and antisocial application to get round the exclusion, describe it in terms which cannot be objected to.

A device for adding zest to sexual activity between male adults could be called a medical appliance of general utility. A machine for projecting obscene political messages on the base of clouds might be claimed as an advertising device.

It's a funny old world.

How Do You Apply For A Patent?

You are at the beginning of a long road which will consume time, effort and money. But the steps forward are clear and, if you move carefully, should be straightforward enough. You fill in a form which describes the invention and deliver it with a fee and a signed application to the Patent Office. The application must outline your idea fully enough to let other people use it after the patent has expired. If you have a better soap, in due course both Joe Smith with his backyard factory and the giant Unilever must be able to manufacture it using the technical information revealed in your application.

You must define the scope of your idea, which you can express as widely as you like. You can only claim something as new in the light of what is already known. A steam engine to power motor cars will not do, unless it is an advance on all previous patents in this area.

There are strict requirements for every aspect of patent application. Consider, for example, the regulations governing the paper you use for the typewritten specification. It must be strong, white, A4 size, typewritten or printed on one side. There should be margins of at least 2.5cm at the left hand side and 2cm at the right. Pages must be easy to turn over, take apart and join together again – for instance, by paper clips, tags or staples – and numbered consecutively.

Applications should include an abstract and drawings. At this first stage, informal sketches will do; even photocopies of drawings from the inventor's daybook are fine, provided that they're clear and legible, and not mixed with extraneous matter. This helps to keep inventors' costs down during the first year, which is helpful at a time when they have no income from the invention. The Patent Office publishes a free booklet, 'How To Prepare a UK Patent Application', which gives all the details.

What Is The Patent Timetable?

Once the Patent Office has logged the application, you are at the 'Patent Applied For' or 'Informal Application' stage. There is then a wait of twelve months before you need to make your next move, during which you can explore your commercial chances. The delay does *not* mean that you have a patent for a year.

At the end of the year three choices are open to you:

- to decide that you do not want to patent, after all. Do nothing. The application lapses and the information in it will remain unpublished.
- to decide that your invention deserves a patent and act on this by carrying on with the application.
- to decide that your invention deserves a patent, but to act on this by filing a fresh application. This tactic lets you improve your invention while you use the first application to establish *priority* for the original concept. In effect, the two applications are melded into one. But you can only do this within twelve months of the filing date of the first application.

A decision to go ahead with an application is usually made if you have employed your preliminary year intelligently and sales prospects look good. The nature of the invention must now be firmly stated. You submit final drawings to replace the original sketches, and – most important – spell out the claim. This reveals what is unique about the idea, what others cannot make without infringing your prospective patent. Unfortunately, because there are few totally original ideas, this usually means that your claim is tightly confined by the past and less likely to make you a millionaire. For instance, any electrical innovation has to better 150 years of existing development. On the other hand, if you have a concept to equal the semi-conductor, you will gain not only a patent but the Nobel Prize.

A Patent Office examiner now begins checking through the records to discover if your invention is really new (you should already be fairly certain of this from your research at the Science Reference and Information Service Library). Between twelve and eighteen months from priority date, the examiner comes up with a search report. You are allowed to change your claim to a certain extent, to work round any problems the examiner has found, such as a similar but not identical past invention.

Your application is published about eighteen months after priority date. Copies become available to anyone, including potential competitors who may wish to steal the idea unless it is fully protected, so you are now fully committed to the patenting process.

Next comes what the Patent Office calls Full Examination, six months from publication date. You are two years down the road from priority, but you now have to wait for the result of your request to be examined. This can take a further six or twelve months. At Full Examination, the application is investigated thoroughly. Does it meet the legal requirements of the Patents Act? Is your description complete and clear? Are previous patents, revealed at the search stage, going to exclude your invention? Is the idea new and not obvious?

You will be sent a report of the results, which can lead to a further period of debate while you offer arguments in reply or change your original specification to accommodate any criticisms.

This process goes on for months. But not for ever. A maximum time of four and a half years is permitted to elapse after the filing date. Then you must either be given your patent or denied it.

How Much Does A Patent Cost?

For a UK patent, the basic charge for filing an application yourself, plus taking it through search and examination stages, is only a few hundred pounds. This assumes you are capable of handling on your own all the procedures I have outlined. This is an unlikely proposition. It is a fair assumption that, if you take professional advice, the cost of establishing one UK patent will be in the region of £2000 (see discussion of patent agents on page 79).

To keep a patent in force, increasing renewal fees have to be paid every year from the fourth year after filing, and you must pay in advance. It's sensiblc to keep the payments up *only* if the patent is earning its keep. If it has earned very little or no money by its fifth birthday, the poor thing is suffering a living death. Don't pay for sentiment – let the patent lapse.

Does A UK Patent Apply Overseas?

No. It has effect only in the UK (and the Isle of Man). If you need wider protection for your idea because it is so promising, there are two ways to do it. The first is through the Patent Co-operation Treaty (PCT), the second is with the European Patent Office (EPO).

The UK Patent Office points out, in an ominous sort of way, the demands of national security. UK residents who have not applied for a national patent *must* obtain clearance before they apply for patents abroad. (The Office also reserves the right to withhold a patent in this country for the same reason.)

No doubt sensitive technology should be kept from countries formerly behind the Iron Curtain. We don't want *them* to know the latest exploding pasta recipe, do we? And weaponry has to be stopped from reaching the Middle East, because the government has such a splendid record in preventing its export.

More seriously, if you have an idea whose dissemination could damage this country's interests, the government may sit on it as it did with Christopher Cockerell's hovercraft and Barnes Wallis's 'bouncing bomb'. The bomb application was filed early in the Second World War, the patent not published until the 1960s.

Unfortunately, any decision to put your invention under wraps is the government's, not yours.

Do You Want A World Patent?

It's a nice thought. One patent application: worldwide protection. Less paperwork, less bother, fewer fees, wide coverage. But there is no planetary scheme available, no method by which you can register a patent in one country and then have it respected worldwide. The great economic blocs of Europe, the USA, Japan and the former USSR (not to mention China) are light-years apart on this issue. They are all too busy guarding their own interests.

But progress *has* been made. By the rules, if you patent in the UK you disclose to the whole world what you are doing. Theoretically, this means you cannot get a patent in any other country. Such a situation would be absurd and international agreements have been signed to sort the problem out.

The Paris Convention

This treaty covers eighty or so countries which have agreed that, if you make a patent application in one of them, you have a further year from first filing to apply in other signatory countries. Disclosure of your innovation doesn't count against you during this time, which is called the priority period.

The Patent Co-operation Treaty (PCT)

This agreement allows one patent application to be filed which covers no fewer than forty-eight countries, apart from those belonging to the former USSR. A single application through the UK Patent Office lists where you want to establish a patent. You pay an increasing fee for up to ten countries; after that, the remainder come at no extra cost.

The application passes to an International Searching Authority, in this case the European Patent Office at The Hague. Application and search report are published after eighteen months by the World Intellectual Property Organization (WIPO) in Geneva, which is a branch of the United Nations.

The advantage of the PCT is that it spares you the trouble of filing dozens of patent applications, and the one you file is in English. It gives the result of the International Search before much needs paying in the way of fees.

Spending decisions about the number of countries where you need protection can be delayed as much as thirty months from priority date, while the search goes on for overseas markets.

After WIPO has given the application the thumbs up, it is sent to whichever PCT nations are nominated on it. If they approve your application, they can then grant you their national patent. The snag is that, at this later national stage, you have to pay the designated countries' scale of fees, plus translation costs – because you must complete the process in the relevant languages.

Overall, the PCT is an expensive way of buying time, although in certain circumstances, it may be worthwhile; a patent agent will advise you on the point.

The European Patent Office

The UK Patent Office continues to issue patents as it has done since 1852, but the European Patent Office (EPO) is really in the driving seat. The British patent system is becoming rapidly less relevant. The UK Patent Office received 75,000 applications a year in its heyday; this has fallen to fewer than 30,000 now that there is a fully functioning Euro-patent system based in The Hague and Munich.

The EPO receives 60,000 requests for filing each year, half from European countries, a quarter from the USA and a fifth from Japan. Altogether it has handed out a quarter of a million patents since it was founded in 1978.

How does the system work? Requirements for patentability are much the same as in the UK. An invention is new if it does not form part of the state of the art; it involves an inventive step if it is not obvious to a skilled person who knows the state of the art. It must be capable of being applied usefully.

You have twelve months from the first filing date of your national patent application to extend it to Europe. This is the priority period of the Paris Convention. The application can be in English, French or German.

You deliver it to the UK Patent Office, which forwards it to

The Hague. There, applications are checked for novelty, to find out what patents in the same area have been granted in the past. Results are published (for a new application, about eighteen months from the first filing date) and then applications go to the European Patent Office in Munich. There they become part of the process of evaluation of all patent applications by a panel of assessors from several nations. It takes just over three years for a patent to be granted, provided no problems occur along the way.

Basically, the EPO issues patents in bundles. You list in your application how many of the seventeen member countries you want covered by your patent – although the EPO can give you a patent for the UK alone, if that is what you want. The Euro-patent is valid for twenty years from the filing date of the application.

The policy is to keep charges roughly in balance with the combined fees payable for separate patents in the UK, France and Germany. So, if you want a patent in more than three West European countries, it will pay you to work through the EPO rather than national patent offices.

There are other gains from the EPO. It saves time to run one patent through one system, instead of having to apply to many different countries. You want to be making and selling your invention, not wasting your efforts in patent offices.

Legal rights are easier to enforce because the EPO patent is the same in every member country. It is a strong patent, having been checked by three international examiners against a background of twenty-seven million records from seventeen subscribing countries.

Finally, there is no point taking out a Europe-wide patent for an invention with a distinctly local value. Patents are expensive!

The material in this chapter and the next two is given for guidance. You should obtain detailed consideration of any problem with patents or their alternatives from a patent agent, a patent specialist solicitor or the Patent Office itself (see pages 221–7 for addresses).

CHAPTER EIGHT

Patents – the Practice

Patenting is reasonably straightforward, in theory. But in practice it's impossible to demonstrate all the traps to avoid. Here are some of the common ones, illustrated by real cases, including my own.

Use the empathy that I have suggested you should develop. Learn from others' experience. Say, 'There but for the grace of God . . .' and resolve not to be caught in that particular way yourself.

Here's an opening question with some nasty backspin on it. Who owns the invention? Are you sure it's you? It may belong to your employer. If you are the director of a company, for instance, you have a duty to look after its general interests and the invention may therefore be the company's property, not yours.

Lower down the chain of command, matters are more complicated, but basically, if the invention is made in the course of ordinary work or duties, it's the company that wins the jackpot. Outside those normal duties, thc rights remain yours. To avoid arguments which may reach the courts, it is wise to have a contract of employment, or a written job description. They leave less room for expensive disagreements.

Case history

When I became a sound technician with the BBC many years ago, my job was to play records – in those days 78s, big 12-inch shellac discs. There was a slick and frenzied record programme, the *Jack Jackson Show* – still fondly recalled by those who listened to it – where you had to pick out the precise points on discs to play the music in. Normally you found the right point on a disc, put the needle there, revolved the disc half a turn backwards and then waited, with the turntable turned off, for

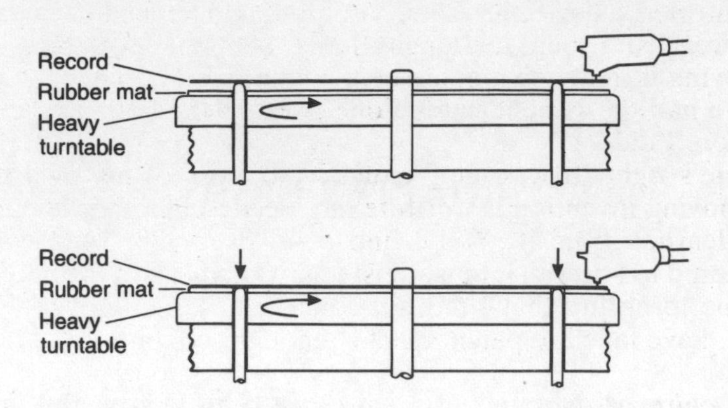

the music cue. When it came, you switched on the turntable and faded up the sound. This brought the music 'on air' with no more than a fraction of a second's error. But it wasn't accurate enough for Jack Jackson's needs.

There had to be a better way. I asked myself: 'Why not?'

Why not use a heavier turntable? Next, lift the record and the already positioned needle off the face of the turntable with four metal pegs rising under the area of disc which overhangs the turntable's edge.

The turntable carries on running freely, while the stationary disc and needle are supported a tiny distance – about a millimetre – above it. When you want to hear the record, you simply push a button, the record and needle descend together on the revolve (which loses no speed because it is heavy and has a lot of inertia), and the entry point on the disc is absolutely precise.

I went to the BBC, my employers, and they said: 'Thank you very much indeed. It's very good.' They installed it and gave me an *ex gratia* payment of two pounds ten shillings (£2.50). I squandered the profits in consolatory draughts of ale.

Get It Patented – Fast

You may feel tempted to fiddle about with your idea, tinkering, improving, polishing, before you take it for patenting. Don't. Get on with your search at the Science Reference and Information Service or one of the provincial patent centres, to make sure that you have a fresh idea. Apply for a patent as soon as you know you are clear.

Case history
Inventor P worked out an idea for removing nasty smells from lavatories, a device he called Ventiloo, and applied for a patent. He recalls: 'I spent too long refining the idea, developing it and proving it in my own home before taking out a patent – others have had the same idea, and one got into the field a year or so before I did.'

He's right. It was me! Actually the story of my own smell-removing invention is worth telling because another lesson can be learned from it. You'll find it on page 128. Mr P will be cheered to know that fate adjusted my handicap in the patenting game to ensure that I did not come out with an advantage. I, too, have failed to patent quickly enough, or correctly.

Case history
Polymers (chemicals with long-chain hydrocarbon molecules) react differently when they are excited by air: some produce a positive, some a negative electric field. I thought of putting two suitable polymers together with a single molecular layer of insulating material between them and blowing air across the sandwich. This produces a sphere of stand-off repulsion. It keeps electrical insulators clean because it stops them picking up dirt, which destroys their insulating power. I thought it would be invaluable anywhere that electricity is transmitted along lines.

I checked the idea, as I keep recommending you to do. It turned out that no one else had come up with it, and the prospects looked good. But before I could patent, the invention had to be tested in a laboratory, as I did not have the high-voltage equipment to do it myself. It was tried out in a lab, and the man who tested it appropriated the idea and took it to the USA. My bad luck. There was nothing I could do: I had no rights that I could legally defend.

There are two points here. The first is that I could have protected myself by applying for a provisional patent. It would have given me a full twelve months to solve testing problems before I needed to go for the full patent.

Initially, you can offer a machine on the basis that it consumes grass and water to produce transport and fertilizer. You may have in mind designing an elephant when you become Patent Applied For; but you may end up after a year's research by changing it to a camel. As long as it achieves the same end, as long as the *concept* remains unaltered and you originally described its scope in broad enough terms – that is, you didn't specifically say that you were thinking of an elephant – it doesn't matter that your final idea is different from the one you first thought of.

It's only in the patent specification you write at the end of the year that you have to spell out details. If things don't work out, you simply let the idea lapse.

The second point is that, because my idea was not only stolen but taken to the USA, even Patent Applied For would not have helped me. It might have led me to the grant of a US patent but I could not have afforded to fight the case in their courts.

Beware of relying on US patents. Many American companies operate on the notion that if they can steal it from you, they would rather do that than buy an honest licence. I have a theory that it's an inheritance from the lawless days of the Frontier. These companies behave on a much more litigious principle than Europeans, saying: 'Take us to court to defend your patent – if you dare.'

The only people who really benefit are the lawyers. So it's useful if you deal with the USA to have a patent lawyer working for you at an early stage. But then the lawyer cuts himself a percentage, and that means a chunk of the equity, the ownership of your company.

Case history

In my back garden I have a model railway track a third of a mile long. I built a new type of model locomotive, a diesel-hydraulic, to run on it. It was an interesting challenge because it had not been done before (see photo section).

Part of the design was a control valve. It gave an unbroken flow of hydraulic drive oil, without stopping the motor when it went from full ahead to full reverse or any position in between. It was a nice idea. You did not have to control the motor; you simply controlled the rate of fluid flow.

What I did not think through properly was the valve's potential for use other than with model trains. I did not bother to patent it; nor did I do my homework to discover if anyone else had already patented it. I was in a position to know how to check these things, and I ought to have known better. The valve is now widely used in industry on automatic lathes, and I have not made a penny from it.

Don't miss other uses for your idea.

What About Euro Patents?

There is no such thing as a Euro patent, though I am guilty of using the expression myself. Strictly speaking, a Euro patent would be a single document issued by the European Patent

Office, valid in every country which is a member of the EPO. The idea has been talked about a lot, but is blocked by the objections of a few countries which see it either as a threat to their sovereignty or to the profitability of their national patent offices.

No, what we are talking about here is the present, highly successful system where the EPO in Munich issues patents which are individually valid in as many countries as the innovator chooses to pay for. The EPO takes the line that anyone offering a new product or invention will already have studied all the literature first. It means *all* the literature, and it is deadly serious about enforcing its rule. This can be infuriating. Why should you fail because of a disregarded and forgotten idea patented in Spanish, in Madrid, in 1891? But you will, if it parallels yours.

I have been privileged to attend the EPO examiners' sessions (an opportunity which is definitely not available to the general public). One aspect which helps the assessments to be fair is that there are three examiners, instead of the one you get in the UK.

Provided that the future of your own innovation is not riding on the result, it can be entertaining to see the way in which assessors from different nations approach an application. Are the French legalistic, the Germans pedantic, the Italians erratic, the British after fair play rather than the literal letter of the law? I'm not telling! I will say that the assessors are impressively well informed.

Whichever you want, national or European, does not affect my general statement that a patent is necessary. The corollary is that you almost certainly need a patent agent as well.

What Is A Patent Agent?

One definition is: an expert engaged at a fee to arrange that you receive a patent if you can show you are entitled to one. Nowadays they are beginning to call themselves 'patent attorneys'. But patent agents are not just people who prepare the drafts of patent applications – in many ways that is the most straightforward of the services they offer their clients, even though it is an indispensable one. They are widely experienced and highly trained specialists who can help with matters ranging far beyond patents: trade marks, designs, copyright, design rights and licensing (see Chapter Nine).

You have to establish a personal relationship, just as you would with a solicitor or an accountant. All patent agents are privy to details about your innermost thinking and secrets, and it is not wise to confide that type of material to people whom you do not trust.

Getting agents to write the patent specification is employing only the smallest part of their skill and knowledge. If you don't feel empathy with them you are using *Ark Royal* as the Woolwich Ferry.

Once the agent is interested in the invention, you have access to advice that more than makes the fees worth paying. In the case of almost every successful invention, the agent becomes a personal friend of the client. There is another side to this. A patent agent has to be chosen with considerable care, to suit you and your problems. These days many agents will give a preliminary consultation free of charge, and it is worth using these, not only to present your innovating problem, but to assess the agent's personality and ability.

Patent agents were originally a branch of the legal profession. They became highly specialised until, in 1891, the Institute of Patent Agents obtained a Royal Charter. These days, those legal origins are reappearing, for some firms of solicitors are starting to take on patent agents and offering patent agency services to their customers.

Generally, though, patent agents work either for companies which require in-house work, or for their own partnerships and agencies. The latter are those you are likely to come across if you are in business as an innovator.

Patent agents have to qualify through two sets of examinations, success in which entitles them to be listed on the official Register of Patent Agents. In addition, most have university degrees or equivalent qualifications. Nearly all agents are also qualified as European Patent Attorneys and can organize the grant of patents through the EPO. The Chartered Institute of Patent Agents (CIPA) controls the profession and imposes a Code of Conduct.

Should You Take Out Your Own Patent?

It's a tempting notion, offering the prospect of saving time and money. If you are a professional engineer working in the appropriate department of a large organization, or for your own innovatory company, go ahead. But most of us are not in that position.

Inventor L comments: *'My invention is in its final patent stage, but if I ever decided to go through this long and expensive procedure again I would be confident of doing most of the work my patent agent has done, myself.'*

Now I know that L is an able mechanic. He has the technical

ability to make out the patent application correctly – and he writes good English. A patent is a legal document, and unless its language is expressed clearly and accurately, you are storing up trouble for yourself. If you are not one of the tiny percentage of the population who can express their thoughts with precision, do not attempt to write a patent yourself.

I recommend inventor L to go it alone. But there are dangers in what he is proposing. The first lies in making the scope of the application too narrow. If it is for a new lawnmower, *don't* say that it is a sheep. After the initial definition, add: 'Examples include domesticated herbivores and grazing animals and mechanical devices.' The word 'examples' is a wonderful solver of tight drafting problems. That keeps the definition broad enough for the next stage of proceedings, where L employs a patent agent to check the final stage of the application.

The agent will find himself bound hand and foot if the document has been too tightly drafted. But if its scope is described in suitably wide terms, he can solve any major problems L cannot cope with. This also holds down L's costs.

To anyone with less experience, less command of the language, or less technical ability than inventor L, I say: hire a patent agent.

How Much Do Agents Cost?

Normally you employ an agent during the Patent Applied For period, the twelve months during which commercial possibilities of the invention can be explored and the patent's claims and abstract must be prepared. I've already mentioned what a quantity and quality of experience is available from good patent agents. They can advise on all aspects of the procedure. Use that advice.

A patent is not cheap. If your innovation is marginally useful, or unlikely to make substantial sums of money, you are advised not to patent at all.

The CIPA itself suggests that preparing and filing a full specification with claims and abstract and paying for the search may cost £1200. More charges will arise at the examination stage (when the examiner reports).

Patent agent B's comment: *'We try to help our private inventors identify methods of exploitation although admittedly this is often frustrated by, quite bluntly, complete lack of commercial reality in these inventors.'*

I totally agree. The honest patent agent simply says: 'I'm sorry. This idea is no go.' If he does, *listen* to him.

Patent agent L's enlargement: *'I once advised a client that he would be well served by the patent system if he let his competitors waste money on trivial patenting, and if he himself only spent money on worthwhile inventions. But he did not accept that advice, much to my profit.'*

Another honest man, whose advice is well worth listening to. I would have been even happier if he had shown the client the door. But if the agent is faced by a client who sees patenting as some kind of virility symbol, a badge of status, there's very little that can be done.

Now let's hear what some inventors think about patent agents. One says: *'There are four parts to patent costs: 1. Patent research. 2. Patent agent fee. 3. Repeat the patent agent fee by giving it to another agent to attack when in draft form. 4. Patent insurance against legal cases. VITAL.'*

I can't agree, on three counts. Patent research is something innovators should do themselves. Even if it isn't their 'scene', they can break the back of the work and leave the agent only the task of assessing a handful of patents which seem similar to their idea. That can save the inventor a lot of money.

Secondly, if the inventor has formed the relationship with the agent I have been suggesting, there should be no need to hire 'another agent' to check the first one's work. I can see that, in a way, the proposal is ingenious, and no doubt there are a few hard cases where it would have been helpful to the innovator, but hard cases make bad law.

Finally, patent insurance is a pricey proposition and insurers are going to want to know that there's a very good chance of success before they agree to back any legal action. If that's the case, why not save the premium?

Another inventor feels that: *'We need patent agents who are prepared to register ideas on a percentage basis rather than demanding large sums of money up front.'*

It won't work, because it forces the patent agent to become part of the inventor's organization. Unless the inventor is prepared to employ him for the rest of his life, he's not likely to accept the proposition. Alternatively, he will ask for a share of the equity rather than the profits. Would you really want to give the invention away like that? Furthermore, the proposal precludes genuinely independent advice. The agent would be less than human if he did not begin considering his own interests as much as the inventor's.

What Do You Do If Things Go Wrong?

'I had an idea which was leaked to competitors. They took out a patent even though I subsequently discovered that the idea was

first patented in 1922 and subsequently in 1953! Therefore the later patent should not have been granted!'

Poor research all round here, I'm afraid. It seems astonishing that not only did the competitors miss the prior patents, but their patent agent *and* the Patent Office did so too. However, it leaves the original inventor in a good position. Since the patent is challengeable, he can go ahead and make the product himself. But he must be absolutely sure there's no real legal barrier to doing so; that the later patent should never have been granted and he can prove it.

Now for a real blockbuster: *'The (very top) firm of London patent agents I used sent a bill for just under £2000 for filing what they called an informal application – way above their original quotation. What was actually filed I subsequently found out contained numerous errors and garbled passages. I believe it was just dictated to a secretary with instructions to file it. I sacked the patent agents. They retorted that it was my fault for not checking the copy. But the letter containing the copy stated that this was what they* had *filed.*

'I went to the firm in the first place because I only wanted one person to handle the matter and not *pass it round the office. Despite assurances from a partner that he would be dealing with it personally it* did *get referred to someone else. Then the partner had a conference with me that left me feeling he was not really on top of it.'*

It hardly seems possible that the bill for an Informal Application (Patent Applied For) could come to £2000. Submit a complaint to the Chartered Institute of Patent Agents, with written evidence. The CIPA has a Code of Conduct.

The inventor himself has made a mistake. He went to a 'very top' firm which is used to handling large-volume business. Perhaps this seems impressive, but it is often not suitable for the private inventor. He would have been better off with a smaller company. The CIPA publishes a free directory of agents, listed geographically, so it's easy to find one not too far away.

Never Sign Anything like This

'I trusted a private organization who claimed and promised to "launch the invention" and find the manufacturer for me. They charged £1700 as an initial fee. They did not do anything other than writing a few letters to some companies.'

There are many ways you can sign away benefits from your invention. There are others where you pay through the nose while the innovation slides quietly into the compost heap of history. Beware!

I am indebted, as with other examples in this book, to Alan Wilcher of Imagineering for the following specimen letter which encapsulates the methods by which an unscrupulous operator can take inventors to the cleaners. Imagine that John Gullible (the inventor) is approached by Reg Greedie of Real Smart Consultants. Greedie offers to see Gullible through the trauma of patenting and merchandising his invention in exchange for a minimal fee. He persuades Gullible to write a letter committing him to the deal.

It ought not to be necessary to spell out what is wrong with this contract. In fact, Alan Wilcher has built so many land-mines into it that it would take me a month of Sundays to pick them all out. Here are a few. Check my comments against the paragraph numbers. The rip-off becomes worse as the contract grinds onwards.

1. 'Any modifications which are based on the same principles or serve the same purpose.'

Gullible is selling himself down the river for ever. One of the arts of patent writing is to devise a new advance every year so that the patent has defence in depth across a period of time. Crafty continual patenting can keep a product protected for many years beyond the first twenty. But who gets the benefit here? Reg Greedie.

2. Greedie gets £850 on the barrel head. For doing what? Nothing.

3. Greedie will certainly advise that the patent has possibilities because he would otherwise get no more cash from the unfortunate Gullible. And how does Gullible know what money Greedie is receiving on his behalf?

4. Gullible has to accept any deal that Greedie offers (even if it is one Greedie has constructed for his own company to exploit the invention). What are 'reasonable out of pocket expenses'? What Greedie judges Gullible can pay is the only possible answer.

5. Gullible is bound hand and foot. If he manages to find a buyer for his idea himself, who reaps the benefit? Reg Greedie.

6. Thirty-five per cent of the *gross* goes to Greedie! Not the nett. Before there are any deductions made for tax, let alone operating expenses which may eat up any income on their own,

J Gullible
1, Carey Street
London W93 2LL

R Greedie Esq.,
Real Smart Consultants
London EC27 9RN

Dear Mr Greedie,

Thank you for sending me your standard agency agreement. I have signed it and returned it to you with this letter.

AGENCY AGREEMENT

1. I have developed and am free to dispose of a new invention which relates to 'Variable Transmission' *and I wish you to act as my exclusive consultant and agent in securing commercial exploitation of that invention, which term includes any modifications which are based on the same principles or serve the same purpose.*

2. As agreed, I enclose herewith a cheque for £850 (at least) *towards your initial expenses. You will carry out an initial assessment of the technical and commercial prospects of the invention and advise me whether there is sufficient interest to warrant proceeding further.*

3. Unless that initial assessment is negative, you will then use your best endeavours to locate licensees or purchasers in the UK and elsewhere for my invention. I hereby authorise you to negotiate on my behalf and to collect all income under such licences and sales on my behalf and you will account to me for that income at agreed intervals.

4. At all times the invention and the income due to me will remain my property and you will seek my approval of any agreement relating to my invention before signature. If I reject an agreement, REAL SMART CONSULTANTS shall have the right to recover all reasonable out of pocket expenses incurred on my behalf.

5. I shall not seek exploitation of the invention except through you nor make any disclosures or commercial proposals relating to my invention without your prior consent and will refer all enquiries relating to my invention to you.

6. In return for your efforts on my behalf, I will pay you 35% (at least) *of the gross value before deduction or addition of taxes of any income in cash or in kind which arises from the exploitation of my invention except as provided in the schedule hereto, even though I may have rejected that deal initially.*

7. Where they are readily available, I will provide documents, models and prototypes of my invention and will demonstrate the invention to prospective clients as requested by you and will follow your reasonable recommendations on seeking patent and other protection for my invention to support any licensing arrangement you may propose.

8. This agreement shall remain in force until terminated by either party on six months written notice. After termination I shall continue to pay you 35% (at least) *of any income I may receive from deals and contracts or approaches arising from your work for me.*

Thank you for putting these words into my mouth and your hand in my pocket.

(Signed)

John Gullible

Greedie takes his cut. Presumably it can come from Gullible's bank account. And Greedie is taking *over one third* even if he makes no effort at all to sell the invention himself.

7. Gullible would have to do all this work with documents, models and prototypes anyway. Why should he pay Greedie? And he still has to organize patent protection himself. Greedie is not offering to do anything for him.

8. Suppose that, eventually, Gullible smells a rat – in this case, a large cage of them. He has no way out. Greedie can carry on in his own sweet way for six months at least. At the end of that time, is Gullible free? No he isn't. Thirty-five per cent is payable for ever.

You may think that this specimen contract is a joke. Well, yes, so it is. But it contains the essence of many swindles which have been perpetrated on innocent innovators. The message is: read the small print of any contract that you are offered by a consultant, manufacturer or distributor. Work out what the legally binding clauses commit you to in practice.

Set aside pleasantries, smiles and nudges which suggest 'we won't treat you unreasonably, we're not like that'. The only good contract is a written one which has been checked by your own lawyer or patent agent. Make sure that you have it.

CHAPTER NINE

Alternatives to Patents

Do you really need a patent? Are there other ways to establish your position? To beat the opposition? Ways which might be less expensive and less time-consuming? Yes, there are.

KNOW-HOW

The best method depends solely on your own initiative and knowledge. It may be a 'wrinkle' about your manufacturing process. Perhaps you have found that clay from one particular source gives a superior result; maybe holding the temperature at 400 degrees for four minutes before increasing it to 800 degrees helps to get rid of impurities. Whatever it is, don't put the information in writing and don't tell anyone about it. Let the competition work it out for themelves.

Another form of defence is attack. Inventor A, who is a Chartered Engineer and holds a Master of Science degree, explains: *'Most of my ideas are in novel ways of designing electronic circuitry so as to do cheaply what is normally done expensively, or to do much more accurately what is normally done crudely or at a much higher cost. I have obtained patents for inventions in the past, but I now feel I get better protection by encapsulating novel circuitry so that it cannot be copied.'*

In the field of electronics, it is only too easy for the opposition to copy a circuit design from a patent and then alter it to avoid infringement lawsuits. To fool copyists (or, more accurately, thieves), add a couple of components which do not belong to the circuit, and solder them in. Make sure that the extra pieces are 'blown' before you install them. Remove all the marked values (which indicate the working status of the component). Then embed the entire circuit in dark resin so that it is very difficult to get at and impossible to see.

The thief cannot look up the design at the Patent Office

because you have not patented its secret. He will be forced to buy your device in order to study it. That's a score to you. But his trouble is only just beginning. Visual inspection tells him that he can't read the values on the components in the circuit; he can't even see the circuit itself. The odds are that he will spend many days painfully scraping away the resin to uncover the components and wiring. What does he find when he succeeds? The layout, yes; component values, no.

Eventually he will have enough information to build a copy of the assembly. But when he applies power to it, *every* component will fail because the 'blown', non-functional, components in your own circuit are fully operational in his. He won't be able to tell which items are booby-trapped and which are normal because they've *all* failed! You've given him a crossword puzzle without any clues.

If there is enough money involved, in the end he will solve the problem. What you have done is gain time to conquer the market. That's the crucial element, commercially speaking.

The deception technique doesn't cost you more than a few pounds. You've lost nothing, because the villain would have stolen your idea anyway if you had patented it. But it's not given to all of us to be Masters of Science like Inventor A, with the specialist knowledge needed to carry out the deception. Nor are we all working in the field of electronics.

What else can you do to protect your invention? There are four methods open to you in addition to patents:

- Registered design. As its name suggests, you file a claim for this at the Design Registry section of the Patent Office.
- Design right. No filing is required for this.
- Copyright. No filing is needed for this, either.
- Trade marks. These do not have to be registered, but you would be unwise to rely on an unregistered mark.

Registered Design

This is a less valuable form of protection for the innovator than a patent, but it is a useful second line of defence. Registered Design gives you a five-year initial ownership of the unique appearance of an article, a period which can be extended to a total of twenty-five years on payment of the appropriate fees. These fees are low compared to those required for a patent – hundreds rather than thousands of pounds – and the whole process of registry can normally be completed in six months. Sounds attractive, doesn't it?

Design is a nebulous concept. The law talks about shape,

configuration, pattern and ornament without defining what exactly these may be. But the idea is not difficult to handle in practice. It concerns the outward appearance of the product. An electric kettle with bulgy sides, instead of the usual straight configuration, would gain registration if no one else makes one. If it has a flame pattern imposed on its outside surface – an idea used to increase its sales – it is likely to be acceptable if that is a unique design.

The difficulty lies in the restrictions placed on registration. If the appearance of your product is determined by the nature or purpose of the invention itself – inevitably, what tends to happen – you cannot register its design. For example, in the common situation where one part interlocks with another on the male–female principle – say, a plug fitting into a socket – their shape is decided by the job they do. Other manufacturers need to be able to use them in order to compete, so design registration will not be accepted.

The next problem is that there is only a second-level advantage in presenting inventions in a way designed to attract buyers' attention – giving them 'eye-appeal'.

You have invented a new bactericidal soap, effective against a variety of skin complaints and ideal for children's soft skins. To crown your marketing effort, you propose to sell the soap moulded and coloured into a shape like an owl. 'Be a Wise Owl – Use BactoSoap.' Provided that no one is already doing this, you may obtain your design registration. But only for the shape. The soap itself remains unprotected, unless you have patented it. So you are back to square one on that aspect of the matter.

There are other complications to registered design:

- As I've already hinted, it must be new. The Design Registry will run a search of its existing designs to check that you are not repeating something that has already been claimed. Fortunately, this restriction is not as tight as it is in the case of patents.

You might be in the business of making gewgaws to sell in steam-railway tourist shops. You decide to offer a series of souvenir products whose outline is based on the shape of Stephenson's *Rocket*. Even though the *Rocket* is internationally known, you could register this design if it has not already been used before in this particular way.

- Ideas which are against law or morality may be excluded by the Registry. I don't know how such concepts can be incorporated into a design – pornographic pictures on the outside of the aforementioned kettle, perhaps?

• Exactly as with patents, the design must not have been disclosed publicly. If the Design Registry spots that you have advertised the product or had it featured in an article, that will be enough to defeat your claim.

The rule is: don't talk and don't display until *after* you have filed your application.

• A final awkwardness is that some categories of design are automatically excluded from design protection, especially works of sculpture, medals, and printed material of a literary or artistic nature.

If, after all this, you feel that design registration will help your product, there is a definite plus to acquiring it. Registration, just like a patent, is a form of ownership. It is a property right. You can make, import, sell or hire out any product which uses the design. You can license other people to use it and draw income from their use of the licence.

You are entitled to prosecute anyone who infringes your design. One of the delightful things about the protection which registration gives you is that it doesn't fall down even if someone copies your design in innocence. A product made this way, looking very like what you have registered, can still infringe your rights.

A sensible precaution is to mark the product and its packaging with the words 'Registered Design No —' because this puts potential infringers on notice about the risk they run. It also improves your chances of obtaining damages in the civil courts.

When you state your claim, be careful to include as many variations of use as possible. Cover all the options. If your design is for a new teapot, do not call it that – 'a container for preparing hot drinks' is more inclusive, and heads off competitors. Try the exercise for yourself, using Owl BactoSoap as your product. Remember: you are protecting the shape, not the soap.

No fewer than 100 nations, including the USA and all western European countries, will recognize a UK design registration if you apply to them within six months of filing in the UK. Over thirty other countries, mostly members of the Commonwealth, will automatically accept a UK design registration as valid (their names are a fascinating list which begins alphabetically with Anguilla and ends with Tuvalu, Uganda and Vanuatu).

Design Right

This is another form of protection for shaped articles (they have to be 3-D, so a carpet pattern or a wallpaper design will not do –

these qualify instead for copyright or, maybe, registered design).

The shape must have something out of the ordinary about it; a familiar, everyday commonplace design cannot gain design right. Nor can items that must fit or match others, which means that spare parts do not qualify.

You do not have to register to acquire the right, but you would be well advised to keep a record of where and when the article was created. One sensible method is to have crucial documents signed and dated by their originator and witnessed by an independent observer.

Defences can be reinforced by packing the relevant papers in a sealed container – and I mean properly sealed, with sealing wax – and posting it to yourself. Check when it comes back to you that the postmark is legible!

The original signed documents should be enough to deter predators and their lawyers. If matters come to a court case where the validity of your papers is challenged, produce the container and ask the court to open it.

That puts you in a sound position to defeat anyone infringing your right or – an eventuality never foreseen by many people – to defend yourself against a similar challenge.

Another sensible move is to mark the product and all sales and promotional literature with the words 'Design Right' followed by the year in which the right became operational. This should discourage not only professional copyists, but those more naïve individuals who do not know that design right exists in the first place.

Design right does *not* give a monopoly position as a registered design or a patent does, although it can be bought or sold in the same way. What it does is let you stop other people copying your product's shape, or trading in unauthorized copies. As usual, you have to protect your own position, which you do by taking a civil action at law.

The privilege lasts for ten years from the date when you first market the article, or fifteen years since its design was created. This is rather weaker than it sounds because, for the final five years, anyone will be entitled to a licence to make and sell products of identical design. Design right is available to citizens of the UK and members of the European Union.

Copyright

A strong and flexible weapon in the hands of innovators, copyright needs no registration at all and offers automatic coverage in a large number of countries. It lets the creators of certain types of work (not necessarily artistic) control the way in which they are used. These works belong to many fields of art or drama, but those of most interest to the technical innovator come under the heading of 'original literary works'.

These do not have to be written to compete with Tolstoy and Dickens, or even achieve the scholarly level of Mills and Boon. A mere list can be entitled to copyright protection; so can a computer programme or a questionnaire, a catalogue or an engineering drawing.

The main test is whether you have applied your own brain to the creation of the document involved. Is the product that of your own hand and mind? Cynics have it that working from one source is plagiarism, working from two is copying, but working from three is original research.

Literary copyright begins at the moment the work is created and lasts fifty years, soon to be extended to seventy inside the European Union, after its creator's death. It's a much longer-lived arrangement than a patent – you have a chance to pass it on to your heirs or your estate – and it's entirely free.

Use of the international © copyright mark, with the name of the holder and the year of publication shown after it, is not essential in the UK. However, if you are not looking actively for trouble, it is sensible to mark any sales or promotion documents in this way.

Many books also carry a statement to the effect that 'the right of — to be identified as author of this work has been asserted by him/her in accordance with the Copyright Designs and Patents Act 1988'. This leaves no one with any excuse for infringing copyright.

It's the same old story with policing your right: your responsibility, your expense. You sue in the civil courts for injunctions and damages.

Where copyright becomes important is in its creative use to protect your invention. Say that you develop a way to assemble a model aeroplane from a kit. This is not patentable, but if you write up the method in a handbook nobody can reprint it without permission. It will also be difficult for any copyists of the aeroplane to produce their own handbook without using some of the descriptive material from yours. If they take too much they will infringe your copyright.

(It's worth noting that photographs are copyright – an important factor when protecting sales brochures and advertising literature.)

The handbook has other defensive applications. One guard against copyists is to make the product so complicated that it can't be used without a couple of months' work with the instruction manual. A number of computer software programmes fall into this category, not accidentally, I suspect. Programs being complex, the handbooks are thick and wordy. In turn, this means that even if the program disks are pirated, copyists will be deterred from photocopying the book. By the time they have sweated over copying several hundred pages, it would have been easier for them to buy the program. Big handbooks deter attempts at multiple piracy and tend to keep it down to the single-user level. And, if caught, copyists of the manual are guilty of breaking the Copyright Act and are liable to prosecution.

Trade Marks

A trade mark is a symbol, made up of words or pictures or both, which builds up loyalty among customers and makes the products using them 'different' from their competitors.

It may seem odd to be discussing trade marks, which are long-term weapons, in the context of new inventions, but you have to look ahead. If your first invention is successful and has a trade mark associated with it, that mark can then be used to help sell your next invention. Provided that your products have integrity, so will your trade marks. What's more, a trade mark is the only form of legal ownership discussed in this chapter whose value grows with the years, instead of shrinking and ultimately disappearing.

The initial fee for obtaining a trade mark, using the services of a member of the Institute of Trade Mark Agents, is around £400. Ownership must be renewed at the end of seven years at a similar price, and then every fourteen years.

When you consider the instant recognizability of famous trade marks – P and O's multicoloured ship's flag, Shell's shell, GEC's curlicued three italic capitals, Ford's name enclosed in a blue oval – the extra value they create is good value indeed.

There is no compulsion to register a trade mark in the UK but it is foolish not to. If you are the owner of an unregistered mark and you find someone else using the same one, you cannot bring a straightforward action for infringement. Instead, you will have to prove that the competitor is 'passing off', as it is called. The first stage in that process is not attacking the opposition but proving your *own* reputation.

The level of reputation needed is very high. If your sign of four coloured balls crossed by the word 'Smitties' has been known in East Anglia for only six months, you won't do well against another 'Smitties' in South Wales. In practical terms, we're talking about having a product which has been on the market nationally for a number of years and achieved substantial sales. Unless you can fairly claim this to be true, it's wiser to avoid the courtroom by electing to register your mark.

Another point is that the process of choosing a trade mark is fraught with difficulty. If it describes the product, it is not acceptable. A manual device for removing wallpaper whose trademark says 'Wallscraper' is not admissible. As many as fifty per cent of proposed trade marks do not pass the test, so cannot gain registration.

Yet companies continue to waste money on turning out trade-marked literature and stamped or marked products *before* they search the Registry, only to discover then that the mark is unacceptable! It's sheer foolishness. For under £100 spent on search fees, they can save money, have peace of mind and know that their proposed trade mark is valid.

If you do achieve registration, competitors are deterred from copying, and various valuable options open up. Licensing is one. I'll return to it later, but a foretaste of its advantages comes with trade marks. They can limit a franchise to a specific area. An example is milk rounds, where the self-employed milkman can only use the trade mark of X's dairy on his float over a named delivery route.

The Trade Mark Registration Act dates back to 1875 and the UK's first registered trade mark incorporates the red triangle and company name, Bass and Co, still used by the Bass brewing company. (The company owns UK trade mark registration number Two as well, and used to own registration number Three.) The triangle was originally a shipping mark. There is a tale that the company owes its niche in history to an employee who spent a miserable night outdoors to make sure of heading the queue the morning the registrar's office first opened.

Trade marks have created their own form of commercial crime, where villains pass off their material as the product of the innovating company by using its mark. There have been many unauthorized uses of the Bass triangle; the company has hundreds of examples, detected over the years. Copies of Parker pens in the Far East pass under such names as P. Arker.

Trade marks can cause problems by passing into the language. Have you ever called a vacuum flask a Thermos, adhesive tape Sellotape or a vacuum cleaner a Hoover? Companies tend to start legal actions against people who use their trade names loosely because if the mark is allowed to become part of common speech, the company can lose its ownership rights. It's sad, because there is no greater compliment to innovation and honesty than having your trade mark taken into the language. Rolls-Royce symbolizes quality; we describe something as the Rolls-Royce of its field to define the ultimate in quality and expense.

The origins of trade marks are worthy of a history in their own right.

• The Birds Eye mark takes its name from Clarence Birdseye, who started out as a biologist working in darkest Labrador. He applied methods of preserving food used by the Inuit (Eskimos) to industrial deep-freezing of peas and beans and a name was born.

• Maxwell House Coffee was first available to the public at an hotel called the Maxwell House in Nashville, Tennessee. Allegedly, Theodore Roosevelt stayed at the hotel and remarked that the coffee was 'good to the last drop' – an endorsement which the coffee company used for nearly a century.

• Henry Heinz christened his foods with the label '57 Varieties' because, although he had around sixty products on the market, he felt that the numbers five and seven combined had a special flavour to them.

• Lemon Hart Rum has no lemon in it; Lemon Hart sold wine and spirits in Cornwall.

• 4711 Eau de Cologne was first sold at an address of this number in Cologne itself as far back as the 1790s.

• Courvoisier is 'The Brandy of Napoleon'. It was christened when British sailors taking the defeated Emperor to exile on St Helena had the chance to sample and massively approve his personal supply of cognac, selected for him by Emmanuel Courvoisier.

There is no magic method of choosing a trade mark – no secret formula, no surefire way to success. What happens is that an innovator chooses a mark from personal experience. It first gains credibility from the worth of the product it represents – not the other way round. It is at the second stage that it becomes immensely valuable, when the good reputation implied by a well-known trade mark is transferred to other products made by the same company.

The downside of registered design or design right, copyright or trade marks is that a determined and well-heeled opponent may find a way around them. There is no such thing as a guarantee against theft. Don't assume that, under any system I have discussed, you are totally protected. Acquire the protection you can afford. But always remember: the only true defence is to sell the invention ahead of the competition, to innovate continually, to sell a better or cheaper product and to keep your mouth shut.

CHAPTER TEN

Into Production

Do you truly want to become a captain of industry? What do you want out of inventing? For a surprising number of people, it is not money or power but the enjoyment of public recognition, or the pleasure of pioneering.

You must decide whether to make your invention yourself or find someone else to do it (and sell it as well). It's important to analyse your own motives. The question is: do you want to run your own company? If you are a bus driver with a new design for central-heating boilers, but happy driving buses, are you willing to change your way of life?

Going ahead with your own invention costs money. You need to find premises and staff. Tools and equipment have to be designed and built. If components are built elsewhere, you may have to pay for the tooling and retain little control over quality and price.

Management of inventions is a skill in its own right. It involves costing, procurement, running a workforce, promotion and sales. There are meetings, accountancy, handling money and dealings with clients. Some people take to it like ducks to water, others quack a little and sink.

• *'I think the basic problem in innovation is that it involves a long chain of events which require different personalities and mental attributes. It is rare that one person can combine all these.'*

Are you one of those persons? Let's consider in detail just one item from that list above: money. How do innovators raise it? The chilling answer is that many fail to do so and their invention goes to the wall. Others succeed only by using their own savings. Leaving aside the experiences of the failures, replies to my survey of innovators give some idea of where the cash comes from:

- *'I raised private money and am now manufacturing the product through subcontracting. Money, or lack of loans, grants etc is the only* real *obstacle when trying to realize one's innovation.'*

- *'I have used my own money i.e. from my own company plus some from me privately.'*

- *'Raising money for an invention is the most difficult thing to do, no matter how good the device. I would not even bother to try nowadays. I always either drop the idea or manufacture it myself.'*

Those comments come from a highly qualified trio of innovators who know what they are talking about. Another group has been able to raise capital in the commercial or public sector:

- *'Initially, I took out a building society loan when I pretended to improve the house* [this is illegal!]. *We currently sell about £400,000 worth a year so can afford to self-fund other new ideas.'*

- *'The bank manager had seen it on* Tomorrow's World, *so had the Business Enterprise people.'*

- *'I was fortunate in that my first two inventions sold well and continue to sell after some twenty years which helped to finance my own small business (and also some failures!). There was no need to raise or borrow money.'*

Suppose that you don't want to mortgage yourself up to the hilt and that you cannot borrow from friends. You have to approach the money market. To most of us, that means talking to the High Street banks.

Bank Finance

Banks want a business plan. This raises many reasonable questions on the legal, financial and marketing sides of what you hope to do, projections of income, cash flow and return on capital. It also touches on your character and motives in starting up in business. You cannot really complain about this – probity and good background *are* important in anybody running a company.

The catch comes with the bank's request for security for the loan. It wants a cast-iron guarantee that if you fold up it will get its money back. This condition has scuppered more new companies than I can count, because the moment any minor financial hiccup occurs, the bank panics and asks for its money, thereby wrecking the business.

Avoid banks! They will be looking for a return of fifteen to twenty per cent on capital in the short term – a target almost impossible to hit.

Inventor H hasn't even been allowed to start borrowing, but the reason is evident in his own comment. He couldn't service the loan if he was given it. He's trapped in a vicious circle: *'Now that a superb prototype has been made, the bank will not grant* any *overdraft facility to enable readily available off-the-shelf standard parts to be procured to make up a batch of machines for sale to the public – not that we could stand the interest rate we would be charged.* C'est la vie!'

Venture Capital

Venture capital and risk capital are not the same thing. Venture capital comes from the big boys, who believe in risk for everyone except themselves and charge the earth for the use of their money. They want your business plan, assurance that your idea will be successful, before it is manufactured or even test-marketed, and most or all of the ownership of the idea. Then you can go away, thank you very much.

- *'Whatever they say, finance houses and banks are only interested in short-term loans and low-risk ventures. It takes approximately seven years to establish a new company from scratch.'*

There are exceptions to the general rule. A company director who devised a range of tools of especial value to the disabled was cleanly bought out: *'I sold the company to Guinness Mahon – a city bank. The products are now sold exclusively by a new company.'* Another exception was a group MD in the plastics industry until he retired: *'I have been able to raise venture capital based on my career and achievements i.e. track record in business. In every case this was conditional on my personal involvement in running the enterprise.'*

The 'seed capital' offered by various government bodies is just venture capital under another name. There's a lesser risk of losing control of your invention, but a much greater chance that there won't be enough money to see you through to the bitter end.

The state has helped a little in the past but that is being phased out. Nor is it likely that any other government is going to help its nationals. There has been a sea-change in the world's political and industrial thinking. It is nothing to do with party politics, more to do with the continued impact of recession. Compare what's happened with the way the Great Crash of 1929 altered the thinking of an entire generation.

Too many good ideas are still leaving the country because they cannot find support at home: *'I tried small British company oil blenders* [about a new design of oil filter] – *then I realized they were going skint. When I approached the Japanese, they said yes, and put up £500,000 for 25 per cent of the company plus marketing and legal and offices etc.'*

Risk Capital

Risk capital is what's available in the betting industry. The investor is prepared to lose in exchange for the chance to make a fortune. But he may be in trouble himself.

• *'Development money from my own resources became a problem, as did maintenance of the patents, along with the patent agents' fees. In the early stages, one substantial lump sum was obtained by sale of an exploitation option to a financial whiz-kid of the time, but his empire collapsed without any exercise of the option.'*

Even for people who call themselves professional inventors, raising money is not easy. Nor are the skills of management, production and selling simple to acquire. Very often an innovator should neither want nor try to be a businessman, but hand the job over to others.

If you own your own company, you should make yourself head of R and D, unless you are well qualified to be managing director. It is usually too difficult to have a clear view about the commercial future of your own ideas. Buy expertise. And fire it if it doesn't come up to scratch.

The alternative to going into business on your own account is to discover a manufacturer. This brings me to the next rule for inventors:

Rule Eight: Find a product champion

If, and it's a big if, you can persuade a manufacturer to take your invention on board, many of your basic problems will be solved (though that's not to say new ones won't appear). The manufacturer has capability in place, so you don't need to create it; he has a labour force, so you don't have to train one;

he has a sales organization; and he has capital to finance production.

Once convinced that your idea is a money-maker, he becomes your *product champion*. Like a medieval knight riding at a tourney for the honour of a lady, he will fight for your cause because it is his as well. But how do you go about persuading him to do battle?

Too many innovators are fools to themselves at this point. Convinced of the intrinsic value of their invention, they neglect all the wiles and professional arts by which they would normally seek to woo a member of the opposite sex. Dressing suitably, conversing amiably, delivering the roses you promise, matter as much in winning over an attractive company as they do with any other love object.

Prepare yourself as for a blind date:

- The invention must look totally professional: no loose wires, no meters strapped on the side of the equipment, no scruffy paintwork. Be sure too that the demonstration works 100 times in every 100.

- The invention should be both functional and aesthetic. Ergonomics matters. In a consumer society, pure design is often a main selling point. Get the colour and shape correct and comfortable. Hire a designer to help you, if need be. Simplify, always simplify.

- Dress smartly. Trousers with the knees bagging out, grease-filled fingernails, and ball-point pens stuck in the outside breast pocket of a sports jacket are not advisable. In fact, a sports jacket isn't advisable either. Wear a suit.

- Polish up your patter. Rehearse what you want to say with the help of your partner or a willing friend. Practise not saying too much – innovators are notorious for blethering enthusiastically. Think yourself into the position of the manufacturer you plan to meet. Use empathy once more. Forget what interests *you*. Think about what will concern him or her.

What problems will your idea create in terms of retooling to produce a new line of goods? What capital investment is likely to be required? Will the sales force need retraining? What should be the market price of your device? Above all, what sales and profits are likely to accrue?

Be aware too, that there are emotional obstacles which concern pride and established ways of thinking.

Defeat NIH

The company's experts may say your innovation doesn't work. It could be because they didn't think of it themselves (Not Invented Here – NIH) but they are not being malicious or obstructive – it's just that your proposal doesn't fit with their own mind-set. They have a pattern of reaction to new ideas, cast in stone by a training from which they cannot break free. NIH is a powerful and corrosive force.

Case history

When I was in the Navy I drove a gunboat in the Mediterranean. The armament on the ship was handled by trained gunners, but I thought we might do better than that. Why not use the ship itself as an aiming device? I got hold of an Italian 20mm cannon which used ammunition similar to our own. It had a fixed degree of elevation, so we could only hit objects at a fixed distance. But it was ideal for attacking land-based targets which cannot move.

We put a sighting ring on the bridge, and a remote-control trigger on the gun, also worked from the bridge. Now we had an extra gun which I could fire by hand by pulling on an adapted bicycle brake lever. All I had to do was swing the head of the ship, and I could traverse a target and knock hell out of it.

It wasn't a real invention, but it served its purpose, and I don't think that the targets we hit worried about its patentable niceties. But when Authority – in this case Admiralty – learned what we were up to, we received strict orders to remove the gun. Allegedly, the ship was not designed to take the extra weapon. The truth was, Admiralty hadn't invented it. It was the first time I had come across the NIH syndrome. As a serving officer, I was not in a position at the time to argue about it.

Rule Nine: Sell yourself as well as the invention

Now that you have prepared yourself to pay suit to a manufacturer, how exactly do you make your overture? The crucial element is the 'cold' letter, introducing your invention. The first example (see page opposite) is such a disaster that you might think nobody could write anything like it.

3, Slippages Terrace
Millennium Street
Chipping Blenniford
Berks SL27 9RN

Relativity plc
Hawking Science Park
Cambridge.

Dear Sirs,

I have invented a machine which will undoubtedly change the world we live in. It uses solar energy to provide a brand-new method of travel and is so highly revolutionery that I am astonished no-one has thought of it before.

I am sure your company will be interested in making this machine. Although it will be expensive to build, its economic advantages are quite remarkable. If I could bring it along to demonstrate to you, which will only take a couple of hours, I am sure you will agree that the invention is unique.

Would you please let me know when would be a convenient time to call?

Yours sincerely,

Bert Wells (inventor)

Don't believe it. I had dozens similar during the years I worked as Unit Co-ordinator for *Tomorrow's World*. How would a manufacturer react to it?

• The letter is not addressed to any individual or department at Relativity plc. Who does Bert Wells think is likely to read it? The chairman, the sales manager? Why not say so? Using the nickname 'Bert' looks unprofessional and adding 'inventor' in brackets is crass.

• Bert doesn't know that letters addressed to 'Dear Sirs' should be signed 'Yours faithfully', not 'Yours sincerely' and he can't spell 'revolutionary'. Is a man with this level of sophistication likely to be an inventor of significance?

The letter seems destined for the circular file beside the office desk, but let's assume the manufacturer gives Bert the benefit of the doubt and reads it once more.

• The invention 'will undoubtedly change the world we live in' – that's what all inventors believe. Where's his evidence? 'It uses solar energy' – oh, dear me.

• Bert wants a couple of hours to run his demonstration. If it takes that long either (a) the demonstration is likely to go wrong, or (b) the invention is too complicated. In either case the answer is no.

• Finally Bert wants the manufacturer to contact *him*. He won't. There aren't enough hours in the day.

Circular file, here Bert's letter comes.

What does the manufacturer make of another letter, also sent by Bert, about the same invention? This is a new Bert, who has taken lessons from Alan Wilcher of Imagineering.

• The first thing to notice is that Bert has become Herbert G. Wells. The full name – it rings a bell but leave that aside – says that he intends to be taken seriously and his position as Managing Director of Universal Travel Ltd reinforces that message.

• Observe that Herbert now has a proper letterhead which, among other things, gives his phone number – a little matter he forgot with his first effort.

UNIVERSAL TRAVEL LTD

Slippages House
Millennium Street
Chipping Blenniford
Berks SL27 9RN
Tel: 0628 999999

30th February 1999

Dr A Einstein, FRS
Managing Director
Relativity plc
Hawking Science Park, Cambridge

Dear Dr Einstein,

COMMERCIAL IN CONFIDENCE

New Product Development – Personal Transport
I have patented an invention whose commercial value may be of interest to you. You will appreciate that I am unwilling to disclose precise details without a signed confidentiality agreement but I should like to bring the following to your attention.

- *Product: rapid personal transport device*
- *Market: international*
- *Features: instantaneous travel – anywhere*
- *Stage: full-scale working model*
- *Protection: provisional patent*

It is my intention to sell the manufacturing and selling rights, preferably on an exclusive basis.

The product fits well with your existing range and I believe that you would be interested in a working demonstration. I shall be in touch next week to find out if I may send you a confidentiality agreement and to make the necessary arrangements.

Yours sincerely,

Herbert G. Wells
Managing Director

• He has also altered 3, Slippages Terrace to Slippages House. It sounds better and nobody from Relativity plc is likely to visit him and discover the hideous architectural truth until after any deal is signed.

This may sound trivial and silly, but it's the body-language of a letter, these almost invisible extra signs, that make Herbert *un homme sérieux*.

• The letter is now addressed to the head of the company (that's where the executive power lies, so never aim lower) and Wells has his name and his qualifications right.

If you don't know the boss's name, ring the company and ask the switchboard operator. She probably won't know his full title, so speak to his secretary. Say that you have to write to Dr Einstein and want to address him correctly. It's easy.

• The letter has been prepared on a word processor but an electric typewriter will do; if you have neither, use a typing bureau. It is laid out in easily digestible chunks – Dr Einstein may have an IQ of 197 but he likes life made easy as much as you do.

• There is a clear brief description of the product, market, features, current stage of the invention and its status as regards protection.

How much has Wells learned about writing an effective letter?

• He *doesn't* give patent or application number(s). If he doesn't tell Relativity plc how many patents he holds on the invention, it will not be inclined to take the chance of stealing it. Relativity might find itself on the wrong end of a law case where it has no defence.

• The letter makes an attractive offer: exclusive manufacturing and selling rights. Wells knows Relativity's product range, so he has researched the market before approaching Dr Einstein. He knows what he is talking about.

• The offer is balanced by the fact that Wells wants the company's initial reaction to his offer of a demonstration within a week. This is an exquisitely polite threat. If Relativity doesn't get to work on his letter immediately, he will take the offer elsewhere.

• Wells is not going to wait for the company to contact him in its own sweet time, if ever. He will ring to discover the state of play.

No one can guarantee that this letter will hit the target. But unless Relativity already has more than enough work, or Dr Einstein is using his MD's position more for support than illumination, Wells stands an excellent chance of being invited to give his demonstration.

Wells has resisted the temptation to make exaggerated claims for his invention. He will follow the same tactic when he meets Einstein and his engineers. There are things he needs to say in explanation; he will say them. But the invention should be able to sell itself on its own merits once he has done so. He knows when to be enthusiastic and when to shut up.

As a precaution, he will take a colleague to any meeting where a deal may be in the offing. The other side is on its own stamping ground, and is able to confront Wells mob-handed. Production managers, marketing experts, lawyers, accountants, can all be used in the nicest possible manner to pressure him into giving way on points where he should not.

It doesn't matter who Wells takes with him. But if he can afford it, a licensing expert, a sharp commercial lawyer or a patent agent is advisable because of their technical knowledge. The point is simply to have someone who reinforces Wells's willpower; someone he can look at and think: If I give way on that point, I'll make an idiot of myself and they will know it. There's a lot to be said for a gigantic but well-dressed friend who says nothing but makes a brief note in a small black book whenever the other side is laying it on heavily.

Once the interview achieves its object, Wells leaves, turning down any offers of a visit to the company bar or the pub. That's where a few incautious words over a drink can wreck hours of self-control and destroy a deal which is all but agreed.

The moment he returns to his own base, he writes a letter of thanks for Dr Einstein's courtesy in seeing him and (the real reason for writing) confirms any decisions arrived at. This reduces the chance that a lapse of memory will cause disagreements; it also keeps the deal's momentum going.

Disclosure In Confidence

The one ticklish matter in the letter is its mention of a confidentiality agreement. This is a document in which Relativity is asked to promise not to reveal any details of what Wells will tell it on the way towards signing a contract. It becomes important when negotiations break down *after* Wells has done the revealing.

Many companies do not like signing such agreements, feeling that they bind them in ways they cannot accept. In this case, it is up to the inventor to decide whether or not to press for the agreement. There's no golden rule to help you here.

• *'Prior to a full patent being granted, an individual may wish to have detailed discussions for licensing the invention, but there is no standard format for a Contract of Disclosure to a third party. I feel that it would be of great benefit to all "weekend inventors" if an agreed Contract of Disclosure could be drawn up between, say, the CBI and the IPI, so that when discussing provisional patents there is an agreed formula on which all parties could work.'*

There have been attempts to construct a standard Contract of Disclosure, but the circumstances surrounding each invention are so complex that it is almost impossible to draw up an umbrella form.

The best effort so far is that proposed by the indefatigable Alan Wilcher. It's as copper-bottomed as his lifetime's experience can make it, but you are advised to have your patent agent check it through and adapt it to your own particular circumstances.

Opposite is the agreement which Wells offered Relativity plc.

Confidentiality Agreement

In consideration of the disclosure of intellectual property know-how and experience by Herbert G. Wells *in respect of:* an invention which permits instantaneous travel to any geographical location or any time in history or the future

To: Relativity plc, Hawking Science Park, Cambridge
(hereinafter called the 'Recipient') the Recipient hereby agrees to hold all drawings disclosures data reports software models samples component parts photographic video patent applications or the like including any oral information relating thereto as 'Confidential Information' given in confidence and not to publish disseminate or use said Confidential Information for any other purpose than in discussing aspects of design development supply manufacturing marketing or assignment of intellectual property rights on terms to be subsequently agreed.

The Recipient will not without written consent from Herbert G. Wells *make any notes sketches drawings photographs or the like. It is further agreed that neither I nor any employee or agent of:*

Relativity plc, Hawking Science Park, Cambridge
will use such information for any purpose other than that stated above without prior written permission of Herbert G. Wells. *Such permission shall not include information which:*

(a) is in the Recipient's knowledge or possession at the time of the disclosure and had been declared as such

(b) is part of the public knowledge or domain at the time of the disclosure

(c) is subsequently received by the Recipient from a third party independently and without binder or secrecy or

(d) subsequently becomes part of the public knowledge or domain through no act or fault of the Recipient such as commercial sale or by the publication of a patent or a technical article.

It is agreed that this Confidentiality agreement is equally binding upon Herbert G. Wells *in respect of the Recipient's commercial secrets and intellectual property rights.*

Should negotiations fail either party will at the written request of the other hand over all relevant documents samples or the like referring to the other disclosed during negotiation.

This agreement shall continue for a period of five (5) years from the date hereof or until said Confidential Information ceases to be confidential as outlined above whichever is earlier.

Signed:............................ Date.............................
for Relativity plc

Signed:............................ Date.............................
for Universal Travel Ltd

CHAPTER ELEVEN

Selling

Relations between innovators and manufacturers resemble love affairs. Courtship sometimes ends in marriage, but often in rebuff. In the latter case, one side – usually the innovator – retires with intensely wounded feelings.

One party often loves more than the other, which is happy to be desired but does not return the emotion as strongly as it is offered. Commonly, the innovator does the wooing and the manufacturer relaxes into the situation. Even if a wedding is finally celebrated – a deal is struck, a contract is signed – the manufacturer thinks more about the offspring of the union (the product) than he does about the partner (the innovator) who helped conceive the baby.

Information passes only one way. The innovator talks incessantly to the manufacturer, but the manufacturer tells the innovator very little. The relationship turns sour through poor communication – just as a marriage fails.

• *'Firms are not interested in new ideas that require expense and development. They do not answer letters, they do not help with criticism and are cowardly not available on the telephone.'*

I have now formed the habit of ringing up and saying, 'Can you tell me yes or no?' Sometimes I don't like the answer but at least I know where I am. If you find that the company won't even talk to you on the telephone, isn't that a message in its own right? Try someone else.

• *'I wrote to many companies (Black and Decker, Lucas, Halfords, generator manufacturers) and to some I gave demonstrations. All were very interested but for a variety of reasons (market too small, too expensive to make . . .) no one took it up.'*

This is what happened to an idea I had, also in the electrical generator market. Generators provide light and power for remote buildings without electricity, such as hill-farms. My own house, buried in the depths of a forest, has two generators, one on permanent stand-by. We are too far from the mains to be connected. But with the expansion of electrical supply, any widespread need for generators has vanished, except for a number of consumers who are not commercially significant. The market today is for replacement generators, supplied to those who already have them and still need them.

You can see why generator builders are always looking for new ideas. But even if they find an invention ingenious, in the final analysis their market is small.

Innovators should study sales potential for their inventions before attempting to launch them on the market.

- *'Getting a firm decision is difficult. Usually engineers haven't the authority to commit their company to expenditure (and very often they haven't the nerve, either!) Higher management are not usually engineers (in Britain).'*

You've been talking to the wrong people. You should start your approach with the managing director; it doesn't matter whether he's an engineer or not. The cold fact is that he is the man who has the power to make decisions – wrong ones included. If you get him interested, he will ask his engineers for a technical assessment and a costing, his marketing people for a sales projection, and *he* will make the final decision, based on what they and you tell him.

Use the empathy I keep referring to. Put yourself in his shoes. What questions does he ask himself? 'Is it worthwhile? Have I got the development money? Are my engineers able to make it? Should we subcontract part of the work? Is the product going to fit with the rest of our range? Where do we position it in the market? Am I going to carry the board with me?' If you can answer those queries on his behalf, as it were, you understand the man and the company you are dealing with. In turn, this strengthens your negotiating position.

- *'I try and sell my invention from a well-prepared drawing with each part explained by notes. I think most sales personnel cannot read drawings – if I receive no response, I scrap the idea and maybe make a fresh start another time.'*

You aren't doing yourself any favours with this approach. Many people can't read drawings; practically nobody will spend time reading notes. Go to any science museum. Cases full of printed information attract very little attention, but there are powered gadgets which operate when you press a button. Whirr-whirr, they go, lights come on, props revolve. The audience stands there entranced and it *learns*. There's nothing like a working model to catch the manufacturer's attention.

• *'I contacted a major engine manufacturer who in the end was dragging his heels for approximately a year, holding up production.'*

Companies often operate on five- to ten-year rolling plans, particularly if they are major players with large production lines. They have to cover much longer periods ahead than exist within lone inventors' experience. Firms can't change everything simply because one plausible idea has swum within their ken. They don't drag their heels deliberately; the costs of change-over can be monumental.

• *'I would like manufacturers to liaise with the inventor in taking prototypes further. I feel that those in authority are too far removed from the practical application to appreciate its benefits.'*

I once sold an invention to a manufacturer with a contractual obligation that my partner should work with him on its construction. We foolishly thought that this solved the problem of liaison. Not only did the manufacturer fail to ask my colleague to come and see him. He actually changed components to save a few pennies, thus guaranteeing that the device would fail when in action.

I learned a lesson from this. The contract should have insisted that tuning and setting up could only be done by us, and that nothing could be sold without our quality-check 'seal of approval'.

• *'I have found it difficult to interest British companies in original ideas. My sales have come in the United States and overseas where they are more receptive to new ideas (and, more importantly, more willing to* pay *for new ideas).'*

Britain has ceased to be a major manufacturing country in the traditional sense. It makes money these days by handling money itself as a raw material. People who handle money are conservative with a small 'c'. When they control the remaining manufacturers – which they do – you aren't going to find many of their firms willing to take chances. American culture is based more on risk-taking. It's easy for Americans to become extremely rich, and it's even easier for them to become extremely poor. Many of them like trying to become rich.

The good news is that you *are* selling your ideas, even if it is abroad. You are bringing money back to this country.

Let Someone Else Take The Risk

One way to avoid aggravation with companies is to switch to licensing. This means that you give someone else the sole right to manufacture and/or sell your product in a defined geographical area. The advantage is that you get your money in the form of regular licence payments, and you don't have to fight everyone else in the market-place. The gain to licensees is that they have a product which has been developed by someone else (who promises to stay off their necks) and an area of sales reserved to them alone.

Licensing practitioners can make technical and market assessments of the chances of success for your new product, find licence or franchise partners, prepare and negotiate agreements and manage the licence during its lifetime. There are good reasons why new hands at innovation should not attempt to negotiate their own licences. It is a subtle area. For instance, are you licensing outwards or inwards i.e. appointing someone to sell your innovation, or buying expertise from someone else? Different techniques are involved.

For those of you who want to understand what is involved, Alan Wilcher of Imagineering offers a simple Manufacturing Licence (opposite). Read, mark, learn and inwardly digest. But don't try to make commercial profit by selling copies of Alan's licence. It is his copyright and he will prosecute infringers.

The licence demonstrates typical clauses – they don't necessarily apply to any one agreement – and it is beautifully clear so that you can understand it without legal knowledge. It may need adapting to your own needs. For instance, if the licence is with someone in another country or operating under a different legal system – Scotland, say, while you live in England – it's useful to have a clause which says: 'This agreement shall be interpreted under the laws of England.' Failure to insert this clause when dealing in Ruritania can end with the discovery that a law passed there in 1762 under the Grand Duke Keingeld messes up your legal rights today.

It's worth explaining one term: the licensing fee mentioned in clauses 6 and 7. This is a fixed payment to encourage the licensee to sell a guaranteed number of Floggits each year; whether he does or not, he still has to pay this sum. It's a great spur to action.

There is a trap in licensing deals which you should look out for. As years pass by, the licensee's thought pattern changes:

Manufacturing Licence for 'The Product'

It is agreed that we have developed a (Product) *for* (doing whatever it does) *and are in possession of drawings, know-how and experience* (and patent application) *relating to* (the Product). *You have expressed the wish to be granted a sole licence to manufacture and sell* (the Products) *in* (give Territory) *which we are willing to grant on the following conditions:*

(1) We hereby grant you a sole licence under our Intellectual Property Rights to manufacture and sell (the Product) *in* (the Territory) *including its components.*

(2) We will provide you with such drawings and specifications and will disclose to you such information relating to the design, construction and operation of (the Product) *as is required for you to manufacture, sell, maintain and service the Product.*

(3) You will treat all drawings, documents costings and other information regarding (the Product) *as strictly confidential and not disclose or make available to third parties without our prior written consent, except as this is strictly necessary for the proper operation of this agreement.*

(4) You will be responsible for all costs of promoting, advertising and demonstrating (the Product) *and will use your best endeavours to expand and develop the markets for* (the Product).

(5) The word Product as used here shall include all improvements and modifications.

(6) A licence fee of . . . per cent will be payable to us based on the gross invoiced price for each product and parts sold by you and for the first . . . units a licensing fee of £ . . . in addition to the licensing fee.

(7) The licence fee (and licensing fee) shall become due on the 1st January, April, July and October of each year in respect of the preceding three months and payment shall be made within 30 days of that due date.

(8) This licence shall remain in force until either party gives the other party 12 months written notice of termination, provided that we can terminate this agreement or convert the licence to a non-sole agreement at any time if you fail to meet the market demand for the product. In the event of termination you will surrender to us all documents, drawings, models and prototypes relating to the product and will continue to respect the confidentiality of information relating to the product.

Year One: 'Floggit's a good product; we'll make a lot of money with it. Thank heaven we were able to buy the licence.'

Year Two: 'We're doing very well; it's a pity Floggit doesn't belong to us.'

Year Three: 'Floggit profits are on the up-and-up for the third year running. They'd be even higher if it weren't for the [word deleted] licensing fees.'

Year Four: 'Isn't there any way we can get out of this licensing deal? We're doing all the work, we're selling Floggit as if there were no tomorrow, and for what? So that this inventor can take our profit.'

Before you know what is happening, the licensee is spending his time and energy trying to dodge making payments. You can solve this problem if you remember that good deals have something in them for *both* sides.

Make the licence run for a fixed period of years, after which the licensee owns the invention in his own territory. Get back some of the income which you are 'losing' by loading the licence fee in the first two or three years and tapering it off afterwards. Do this, and the licence holder will pay up happily. He has every incentive to work hard at selling Floggit because the more he works, the more valuable the property will be when ownership changes hands. If you don't, it will cost you a fortune to police the agreement, and the licensee can rip you off – because the man on the spot has control.

The exception to this suggestion comes if you have reason to suppose that Floggit will be a big seller for ever. In this case, you'd be barmy to give your ownership away. But you can still keep the licensee sweet by putting your fees on a reducing balance: the more Floggit he sells, the less he has to pay per unit.

Whatever happens, never sign a licence agreement where you part with your rights for a percentage of the profit. It is only too easy for a company not to show a profit, even after it has sold 100,000 units of Floggit. This is called creative accountancy and is the cause of much wailing and gnashing of teeth among those created against. So do have any agreement checked by your patent or licensing agent before you swing into action.

Publicity And Marketing

Now let's assume you decide to make and sell Floggit yourself. No outsiders! You will keep everything under unified control. The Floggit production line is nearly ready to go. What's needed now are sales and, to sell, you need literature and advertising. A few simple rules can help.

• Build publicity around actual performance, not claims which the buyer or manufacturer might find difficult to verify. When quoting performance figures of any variety (whether they be revs per minute or number of shirts cleaned in an hour) use proved and independently produced information.

• Make sure that statutory authorities give you the thumbs up. If you design a new fire escape, obtain the approval of the Royal Society for the Prevention of Accidents, as well as conformation to BSI and other national standards.

• Get the most powerful recommendation you can, which is often the cheapest: that of the satisfied user. If that user has tested Floggit, he will advertise it himself to his friends. If he is well known, you may even be able to persuade him to endorse Floggit publicly – for a fee, of course.

• Honesty pays off.

Writing Advertising Copy

Whether you are writing a sales promotion leaflet or designing a press advertisement, the same general rules apply. (I shan't discuss television advertising here, because by the time you are rich enough to afford it, you will have advertising agencies sniffing around begging to work for you.)

• Keep it simple. Avoid long words, unusual words and above all, jokes. Consumers may laugh when they see your witty approach to selling, but they don't buy from comedians.

• Sell to the consumer, not to yourself. You are delighted by the way you solved the technical problems posed by Floggit. Surely buyers will be equally gripped by your ingenuity? Wrong. Consumers want to know what an invention does for them. They couldn't care less how skilfully the sanding head of a power drill is put on a flexible mount. What *does* interest them is that the sander makes it possible to strip paintwork without leaving burns and arc marks all over the underlying wood.

No teenager thought for a moment about the method by which Mickey Munoz attached roller-skate wheels to the underside of a surfboard. But millions of them loved the way skateboards transferred the freedom of the sea's breakers to arid inner-city streets.

If the product is a fairly mundane object, needed to achieve a desirable end, learn from the airlines. They operate vehicles which many of their customers regard as aerial coffins equipped with liberal supplies of alcohol and rubber chicken. What do clever airlines do? They advertise *destinations*. Advertise what the product achieves, not what it is.

• Exploit certain words in the headlines. There are many which are known to work well in drawing the reader's eye. 'New' and 'now' are good and can be used because the product is innovative. 'Announcing' and 'Introducing' fit equally well. Then there are 'Easy', 'Revolutionary', 'Quick', 'Magic', 'Compare'. And don't forget 'Unique', 'Opportunity', 'Health', 'Sex', 'Safety'. Finally, there's the magic three-letter word 'YOU'. You are always interested in reading about you – aren't you?

The headline should tell the product's story on its own, because more people read headlines than the body copy. Advertising agencies working for large companies use this method all the time:

'Does a business package that does the business have to cost a fortune?' Dell Computers.

'The Guaranteed Stock Market Bond. Heads you win. Tails you can't lose.' The Trustee Savings Bank.

'New Season Fashions At Exceptional Value.' Debenhams.

Don't use negatives. 'No More Arthritis With Floggit' will enter the memory of half the population as meaning that Floggit gives them arthritis.

• Use artists' drawings, not photographs. Photos rarely pinpoint the essence of a new idea; in fact they often let it down. A good drawing can lead the reader's eye in the right direction, and written information can be superimposed without a clash between picture and words.

• Lots of explanatory text is OK. There's a general impression that advertising copy has to be short. That's wrong. Well-written stories (factual, not imaginative) draw readers.

• Keep the message cool, clear and factual. Avoid hype. Eschew dishonesty (apart from the fact that you can be prosecuted, it doesn't make for long-term survival in business).

Press Publicity

Asking the press to publicize your work is a weapon that may backfire, so take extreme care. How will your lab or workshop appear? What will the product look like? Publicity can give away vital information to your competitors, and even forfeit your patentability.

As a general rule, you will get straightforward coverage in a local weekly paper – it has too much invested in its readers' goodwill to upset them gratuitously. Regional dailies such as the *Birmingham Post*, the *Western Morning News* and the *Yorkshire Post* are also honest and reasonable.

Once you come under the scrutiny of Fleet Street, it's time to watch out. When you are talking to a specialist – say, the technology editor or the business editor – there is little problem. The specialist may tear your invention to shreds, but will do so with fair arguments. It's general reporters you have to keep an eye on. You have no control over what they may make of your life's work. To many journalists 'inventor' equals 'nutter in a garden shed'. Unfortunately this stereotype is partly based on fact but, equally, many general reporters cannot tell a cotter pin from a spokeshave. Whatever their disqualifications, they have the power to publish an ill-reasoned account of your ideas.

Even if you are lucky enough to have a sensible story written about you, behind the reporter stands the sub-editor who prepares that story for publication. His views of invention are not affected by anything as basic as knowing the facts. He can change a favourable story by altering its angle in the subbing, or by adding a knocking headline. Imagine what could have been done with Marconi's invention of wireless telegraphy – 'Italian Claims He Can Receive Signals From Space.' Factually almost accurate – but what a false impression it creates!

Programmes such as *Tomorrow's World* are a good bet. Television is such an expensive medium that producers can't waste time transmitting items about inventions that don't work. Their criticism is expressed by not putting the story on the air in the first place. But beware of programmes of the chat-show ilk. These are often compèred by smooth clever-dicks from Oxbridge – personally incapable of changing a lightbulb without a PA to help them – whose only notion of invention is that it is good for a few laughs.

The worst of them write jokey books with titles like *Patent Rubbish* and *Completely Loony Inventions*. They are to be avoided unless the interview fees they offer make the insult of their attitude bearable.

Why Exhibit?

Exhibitions are very costly but difficult to resist because of the thrill of meeting other inventors and the exposure they can give – not to mention the ego trip. Unfortunately, such events expose you to competitors, who crowd round while they photograph the product, grill you about specifications and take your literature for closer study.

In my experience, general 'fairs' and 'shows' are not much use; but trade exhibitions are. The people who attend them are all in the right line of business – *your* line of business. Trade exhibitions are useful for finding licensing opportunities. Only go to them when you are ready to manufacture, when the product is flawless. And don't trust a salesman to represent you. Be there yourself. A salesman may claim too much – to your later grief; or he may undersell because he does not understand the product's possibilities. Even if you are not God's gift as a salesman yourself, it is always impressive if the man on the stand can call you forward in response to an inquiry, with the words 'Our technical director can give you the answer to that' – and you can.

Marketing

The first consideration is to get the price right. Price is an area that should be broadly determined at the prototype stage. It does not depend on the volume of sales, as is popularly believed; rather the opposite. The price of the product is set by its value to the consumer, not the cost of production. If the first is lower than the second, don't begin to think about selling. Follow the example of Alan Sugar, who sets in his mind what he thinks shoppers will pay for a product, then ruthlessly argues down the figures his suppliers quote him until that price is met.

An acquaintance of mine wanted to publish a history of his village. He decided that it should be high-quality, with expensive paper and many photographs; it had to be hardback and have a striking, well-designed jacket. From a careful study of the bookshops, he estimated that the highest price buyers would pay for a copy was £13.95. The cheapest quote he obtained for producing it as a complete package left him no profit. He knew he couldn't increase the price of the book without putting sales into a catastrophic nosedive.

By splitting production into several sections – typesetting, block making, cover design, printing and so on, with different firms quoting for each part of the job – and doing much of the work himself, he eventually reduced costs so much that he was able to make many thousands of pounds profit for charity.

Every innovator should follow a cost-cutting approach.

When setting a price, allow for the margin that the retailer will want. To the novice this can appear appallingly high: between a third and a half of the wholesale price. And then there is VAT! A product can easily end up giving the wholesaler i.e. innovator, a little over a pound profit in every ten. Even this margin comes under attack when the shopkeeper asks for a discount for quantity or for prompt settlement.

This is why many innovators end up taking the mail-order route. Costs are under better control and less has to be conceded to middle men. Mail order has another advantage: you can code each advertisement according to the publication it appears in. So when people clip the coupon, their returns tell you which publication is selling the most product. At the next stage of research, different advertisements are run in publications which have proved to have equal effectiveness. The advert which draws the highest number of orders is the best.

Mail order is the only form of advertising where you can work out exactly how effective your campaign has been. For the rest, Lord Lever's remark of some seventy years ago is still true: 'I know that half the money I spend on advertising is wasted, but I don't know which half.'

• *'I made a wrong decision on how to market the product. Direct sales of complete systems rather than going through retail outlets would have been a lot better.'*

Marketing is a skill. Wrong sales decisions are often made because the inventor has no training as a marketing man. What's more, mail order may have been the right decision in this case, but not others. And it may have been the right answer last year but not this year. Situations change.

Find an adviser in the marketing world *before* you launch the product. If you are already in business, The Chartered Institute of Marketing operates a Managed Business Consultancy Service (see page 230). If you are only just starting up, friends can often put you in touch with experts who may be willing to help on an *ad hoc* basis. Hire them for a day's brainstorming or spend an evening in the pub, and talk it over. But get a quotation first and pay them properly.

Good ideas still require active selling. Trained staff and ideas are needed.

• *'My problem is trying to get sales staff to represent the true value of the machines and having agents trained properly to further the aims of the patent.'*

There is no problem getting sales staff. Their competence depends on you, the innovator. Employ people you like, people you can trust, and then train them and let them loose. Your position is no different from that of any other manager. Motivate your staff, and then test them by playing devil's advocate.

Say: 'This product's no good, I don't believe in it or your claims for it. Prove to me that it is worthwhile.' You know the product's weaknesses (as its inventor, you have every reason to know); if the sales staff are up to the standard you've set, they'll still be able to sell the product to you, its most knowledgeable critic.

• *'There's a problem linking production to sales. They seldom appreciate each other's problems, and neither usually appreciates the economics involved for profit!'*

'A problem linking production to sales.' That's what every company faces. Sales and production never understand each other's problems, though they ought to. Get the two groups together to discuss them. Japanese companies, and some American ones, go further. They send their engineers out on the road to sell, and they make their salesmen work in the engineers' office. The head of Sony claims that this experience was the beginning of his success in business.

CHAPTER TWELVE

How Things Go Wrong – Part Two

One of my worst nightmares is of the people or firms I negotiate with stealing my ideas. Or products being copied by companies which appear from nowhere, like birds of prey stooping on unwary rabbits, the moment they smell profit in the air. The more successful the invention, the faster predators gather.

Dishonesty

Ron Hickman had one of the greatest inventing successes of modern times with the Workmate. On the remote chance that there exists someone, somewhere in the world, who has not heard of the Workmate, it is best described as a combination saw-bench cum vice cum work-surface cum stepladder which can be folded and hung out of the way after use.

With the profits from his worldwide sales, Ron (a design engineer) planned and built his own house in Jersey, an Aladdin's cave of delights, full of technical wizardry. One room holds his personal Black Museum, full of illegal copies of the Workmate.

The first major challenge to Ron's patents came from a company that arranged sales of Workmate look-alikes through mail-order companies. It was a potentially disastrous situation. Not only might mail orders soak up the major percentage of his market but, if Ron could not protect the patents, a legal domino effect might destroy them across the rest of the world as well.

One of the things that made all the difference to Ron's successful patent defence was that he was supported by Black and Decker, the Workmate's manufacturers. Ron had a product champion.

Sears Roebuck – the giant American mail-order firm – marketed another imitation. To add insult to injury, Ron had already offered it the Workmate, and had it turned down. Now, here came Sears Roebuck with a similar device under the transparent disguise of Workbuddy. Ron and Black and Decker won this battle, too.

The most extraordinary episode was that of the Kinso, a copy of the Workmate so perfect that Black and Decker itself couldn't tell if it had been made by one of its own factories. When the legal case was won, Black and Decker took over all the stock, re-badged them and sold them as Workmates!

If this can happen to Ron Hickman, it can happen to anyone. Certainly some of the innovators who responded to my survey have been bitten. Their letters seethe with suppressed outrage.

- *'Don't let managing directors into the workshop who notice other ideas.'*

What a picture this conjures up! The innovator demonstrating Idea A, all unknowing, while a swivel-eyed visitor concentrates on the three-quarters-finished prototype of Idea B clamped to the workbench in the corner. The apparently innocent request for a cup of tea; the panther-like bound to the bench the moment the door closes behind the inventor's back; the rapid use of a tiny Minox camera pulled from the waistcoat pocket. It's got a funny side to it, but I *do* sympathize and agree.

Guard not just against managing directors, but any visitor. The most valuable property that innovators possess is inside their heads; the next most valuable is the drawings, working models and prototypes which express their ideas. Giant enterprises employ uniformed guards and dogs to watch over their premises. The least *you* can do is keep confidential papers in a safe and lock away working models.

There's computer security, too. Code a hard disk to prevent access. Lock floppies in the safe with the confidential papers and take the back-ups home each night. Vital information is lost as easily in a fire as it is to burglary. During the day, drape a sheet over what you want to conceal. That's what the Navy does with ship movement charts. It's primitive, but effective.

- *'As soon as I confided my patent specification to the company (before any licence agreement had been signed) they copied a part of the technology, proceeded to make a quantity of products and offered such parts for sale to the public without my knowledge or permission and of course no royalty fees.'*

If the only result of this book were to stop any more private inventors handing over their patent specifications, my declining years would be made happy. In negotiation, you don't have to reveal any more than the working details of the invention. The nature of your patent claim remains secret from everybody for eighteen months, precisely so that you can improve the specification and, in confidence, try to find a commercial home for it. If you disclose your legal claim, some people find it more than flesh and blood can bear to miss a chance of gaining the invention's benefits for nothing.

This letter says 'they copied a part of the technology' and clearly what they did was to steal only that part of the invention where they were safe to do so. This they could not have done without the information the inventor gave them. If you protect yourself the way I suggest, the company may coyly hint that it can take the information already given to it, adapt it a bit and put out the invention itself unless you reveal the patent specification. Ignore this. It's a bluff. They don't know how many applications you have submitted, how wide-ranging your claims are, or how many of them exist.

Challenge them to go ahead and infringe your patents. Say that you will be delighted to sue them for every penny they possess – it's an easier way of making money than by licensing or selling. They then find themselves facing a risk that no normal firm will be prepared to take. Take, and hold, the initiative.

• *'In theory, the small inventor is protected by a patent, but in practice large companies pick up any inventions which may be useful to them with little thought for the inventor, and no intention of paying for the use of the invention. Should the inventor attempt to take the large company to court for infringement, it uses delaying tactics repeatedly, asking for postponements to obtain more evidence. This costs nothing, since it has legal departments in existence anyway.*

'The small inventor, however, has to employ a solicitor and counsel. Every delay costs him perhaps thousands of pounds. Sooner or later, he will cut his losses and drop the action.'

A patent doesn't protect the inventor – it only proves his or her position as originator. The *inventor* has to protect the patent. If there aren't the means to do so (and it's true that a big company can outspend the small man in the courts), like Ron Hickman you have to find a product champion who will fight the battle for you because he is fighting his own battle at the same time.

I repeat the advice given in answer to the previous letter. In the opening eighteen-month period, the big company hand-off works solely when the company has obtained details of the patent application(s). Only the inventor can have given them away. Don't do it.

• *'I trusted people with prototypes for evaluation purposes only to have them returned broken and obviously looked at for plagiaristic purposes.'*

People are extremely careless with prototypes. It's most upsetting, but it seems to be a fact of life. 'Plagiaristic?' It could be, but there's a danger of becoming paranoid. Given a prototype, you too would take it apart. You would want to make sure how it works. If you ask for an evaluation, you must expect close inspection.

• *'I took a person's word as his bond.'*

Every successful con merchant is believable, likeable, charming, and utterly 'honest', otherwise he wouldn't be successful. He doesn't wear a striped jersey or carry a bag marked 'swag'. Honeyed words are his jemmy. Sam Goldwyn summed it up: verbal contracts aren't worth the paper they're written on.

There's only one country, as far as I know, where you can actually win a court case based on a person's word, and that's Scotland. There, a handshake does mean a firm contract. What a sensible system. Why isn't it adopted everywhere else? Get it in writing.

Bad Timing

Timing is largely outside the innovator's control. If an invention fails, it may be because current technology is not sufficiently advanced or because society does not yet demand it. In 1937 Alec Reeves patented a method to improve the recording and transmission of sound. Sound takes the form of a continuous wave. Reeves chopped the wave into separate bits, which he then replayed rapidly enough for the human ear to assume that it was listening to a wave, but the quality of the sound was enhanced.

Unfortunately, Reeves's system failed because it relied on thermionic valves. They generated too much heat for his idea to be workable. Once cold semi-conductors came into use in the 1970s, digital sound – which, in essence, Reeves was proposing – became possible. It is increasingly common today. But Reeves's patent expired in 1961.

Case history

My design for a smell-free lavatory was a classic case of inventing at the wrong time. I noticed that in our house in Guildford, nasty odours from the loo blew everywhere when the wind was in the wrong direction. The answer was to use more effectively the extractor fan already installed in the lavatory.

For flushing, a sparge pipe drops water from the cistern through the rim of the lavatory bowl. I ran an extra tube from this pipe to a powered extractor fan high up on the wall, so that the fan sucks air out of the lavatory bowl through its rim. When the user sits on the loo, the fleshy behind forms a seal – a gasket – on the seat. Air enters the bowl through the gap under the seat's rim. A concentrated air flow created by the extractor fan purges the bowl of unpleasant smells, and pushes them straight outside the house to the open air. It works splendidly.

I thought there was a real need for the idea and I found no competing patent. So I went through the expensive process of patenting. But I could not sell the device at all in the UK because, I now know, my idea was ahead of its time. Society was not yet delicate enough to find the idea valuable to civilized living.

Attitudes were different in continental Europe. In the mid 1980s, at a time when the Iron Curtain still existed, I happened on a remote Austrian village on the border with Hungary. The village had suffered badly from the loss of its cross-border connections to Hungary: it was really run down and poor. But, in the lavatory of the village pub, there was my invention! I was selling at the wrong time (or maybe in the wrong place).

There is a pleasant postscript to the story. A well-known English firm of sanitary-ware makers took up my idea after its patent expired. Legally speaking, they didn't owe me anything. But they made me an *ex gratia* payment in the form of a wonderful bathroom which I still enjoy every day of the week. There *are* honest and honourable people about.

Case history

Ron Hickman invented a device with an internal light source for inspecting the underside or inside of anything that needs such inspection. The Peeping Tom, as he calls it, unfolds in such a way that its small viewing head can be shoved under cars to inspect a suspected faulty exhaust system, or into pipes to find holes, or anywhere else where it is difficult for a human eye to gain access. (One of the problems with the human eye is that it lives in a human head, which may be so large that the eye cannot get where it needs to be to make an observation.)

Ron couldn't sell his 'goes-under' mirror, partly because it was complex in design – he found it difficult to get the wiring up its folding stem – but more because he could not make its fibre optics work. He was ahead of his time. More recently, technology has caught up and an improved design has been tried out by the Army and the Home Office to detect terrorist bombs under vehicles.

Marketing Errors

As this chapter seems to be largely about Ron Hickman, and because he is refreshingly open about admitting that he is not infallible, he will not object if I tell a story, for other inventors' benefit, that shows how a good product still needs to be directed accurately towards buyers.

Case history

Ron developed a new type of potty for babies. He noticed that when they stand up after using the pot, it tends to stick to their behinds and then fall off, with disastrous results. The solution is obvious, isn't it? Here's another exercise for the trainee inventor. How would *you* solve the problem?

Ron's answer was to put a flat rim at the bottom of the potty. The baby's feet naturally rest on it, so that when it stands up, the potty is held firmly to the ground. Problem solved.

However, the idea went wrong at the marketing stage – Ron isn't sure how. Sales were abysmal. Mothers didn't understand the purpose of the invention. One of them commented to him: 'Oh! Isn't this lovely? The little darling's got somewhere to put his feet!'

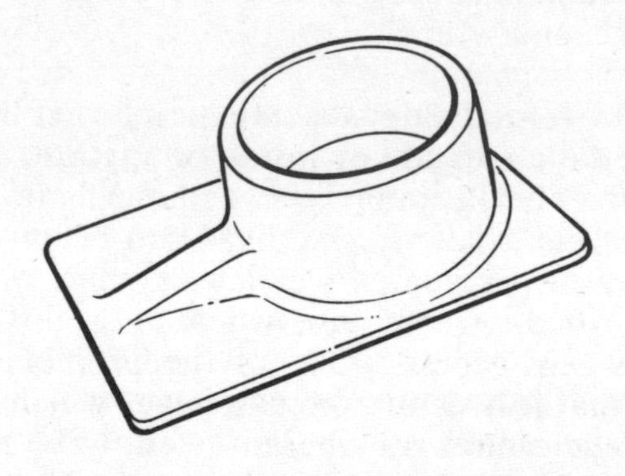

Changes In The Law

Government has gone legislation-mad in recent years. Infringing regulations governing health and safety which our parents would have laughed at as mollycoddling carries large financial penalties. In addition, European Union rules cover every aspect of our lives, and they are viciously administered in the UK by groups of crazed bureaucrats who seem determined to make the EU's name stink in the nostrils of ordinary citizens.

Case history

This is more of a tale I partly narrated in Chapter Eleven. People without mains electricity make it for themselves with generators. But it is a wasteful way of working. Turn on a single light bulb in the early evening and the generator has to clank into full action. The expense in unnecessary fuel costs is horrendous; you may be running a three-kilowatt generator to light a sixty-watt bulb. Also, every time you start the machine, you wear it out a little more.

I decided that the answer was to have a linked array of batteries – car batteries – which deliver power at a low level and run lights without calling on the generator. I designed a logic circuit which 'tells' the generator to switch on only when it is really needed for hard work (when an electric fire is turned on, for example). At that time, the generator also recharges the batteries so that they are ready the next time you call upon them.

Neat and effective, the system has been working in my home for nearly twenty years and I have sold the product commercially. But I face a legal problem. The Consumer Protection Act says that if the circuit happens to fail – say with someone in an iron lung, or if a user happens to get a shock – I could be liable for damages. I cannot afford to pay for insurance to cover this risk. Anyone who makes generators carries that risk anyway, so I am now looking for a manufacturing licensee, someone to whom I can lease the idea. They would make it and sell it and I would draw a royalty. If this does not turn out to be possible, I will be driven out of business.

Lack Of Venture Capital

In Chapter Four I introduced Mr Riad Roomi, the inventor of Ready-Stitch, a surgical innovation which makes it possible for a surgeon to repair wounds without leaving operating scars. The proposal was hailed by professional colleagues and won every innovation award for which it was entered. The future looked bright. What has happened since then? Mr Roomi tells the story in his own words:

'I was approached by a leading surgical suture-material company who showed a very keen interest in marketing Ready-Stitch. That was just after I had won the awards. At that stage, a reputable product design firm were involved in the technical designs for the product. They suggested that the best way to attach the threads to the plaster was to weave a tape to hold the threads together, and then sandwich that tape between two layers of adhesive surgical plaster.

'This route meant engaging two technical companies to develop the product, weaving and adhesive-tape manufacturers. The weaving company worked very hard on solving the problem but unfortunately they couldn't. Therefore they engaged the expertise of another company in Switzerland. The latter produced a miniature sample but said that in order to make a larger one, the real one, a new machine needed to be built which would cost between £60,000 and £80,000. The weaving manufacturers were prepared to invest in building this machine, on the strength of a firm order or contract of some sort.

'Similarly, the adhesive tape manufacturers were keenly interested and produced miniature samples and were happy to enter into agreement.

'Negotiations continued with the potential licensee, who made it clear that they had no interest in manufacturing the product themselves and therefore would require the developing companies to manufacture it for them. Negotiations continued for nearly a year and a substantial sum was provisionally agreed. However, one big question began to surface, which was, do we have a fully developed product before we sign the licence? The answer was clearly no.

'It was a chicken-and-egg situation. The manufacturing companies would only invest in building the machine if they could see a contract, and the licensees would only issue the contract if they could see a fully developed product. Negotiations suddenly collapsed.

'Despite my loyalty to this country, I had no option but to go to the United States. I now have a contract with a reputable joint-venture company in Boulder, Colorado, which I am going to sign fairly soon. This contract is to licence the company to carry out the R and D work on Ready-Stitch and market, if they see fit. I am glad to say that I can still see a light at the end of the tunnel.'

Let's try to be philosophical about this. The important thing for accident victims and hospital patients is that Mr Roomi's idea goes into production *somewhere*. It's also the case that if Ready-Stitch is ultimately profitable, the licensing fees come back to the country where the originator lives. But I grind my teeth to think that, for the lack of less than £100,000, Ready-Stitch is joining the invention drain. And maybe its innovator will follow, despite his avowed wish to be loyal to the UK.

The Inventor Problem

There are other difficulties, such as lack of help in developing ideas, disbelief and jealousy. A major problem with inventions is inventors themselves. They can have wildly over-optimistic ideas about the future of their invention. Many lack knowledge of materials and markets. More find it impossible to put themselves in the customer's or the manufacturer's shoes.

Ninety per cent of inventors who go to exhibitions (by which time they really ought to be ready to offer a prototype or pre-production model) suffer from not having done enough homework. They don't read, they don't investigate, they don't check. They don't look at the market.

- *'I believed that a proven concept will sell itself in the market.'*

What is 'a proven concept?' Who is the judge of 'proven?'

I suspect that there is a confusion here between *Identify the problem* and *Meet a need*. The product may indeed solve a problem brilliantly but that doesn't mean anyone will want to make it, distribute it or – the crunch – buy it.

- *'I spent too much time exploring good ideas which did not have a high enough unit cost to be significantly rewarding/profitable and were far too expensive to develop and launch.'*

This is an honest admission. Many other innovators commit the same error. One constructive thought is that there's a market which loves items with a low unit cost. It wants them in millions. That's the 'give-away' business. Many companies promote their sales with free products which they buy from 'give-away' manufacturers. The inventor can make them, or license them for manufacture, but the point is that they are cheap and cheerful and sold *en masse*. The fact that profit is 50p an item does not matter so much when a mail-order company takes 100,000 of them.

- *'I expected others to be instantly impressed and relieve me of any further (hard) work!'*

I recognize the subject of this self-mocking comment. All I have to do is look in the mirror! We tend to fall in love with our own ideas. I try to follow advice I was given by my father: 'Put it down, stand away from it and come back later.' Returning, you find you have shed that blind, loving relationship with the idea and can see where improvements should be made. Spend the intervening time dancing, partying, skiing, breaking a leg, anything. Get away into something different. It almost always works – and if it doesn't, you'll have had a good time anyway!

- *'My major mistake was underestimating the full cost of patenting. This has exceeded our patent agents' original estimates by more than 30%.'*

It is not unusual for estimates to be exceeded. As with building a bridge, or the Channel Tunnel or any single engineering structure, it is difficult not to underestimate.

There are always hidden factors which you don't expect because you aren't dealing with a production line which has been running for a year, with its hiccups all sorted out and understood. A patent is a one-off, with the problems that one-offs involve. The wise innovator builds a margin into the budget to allow for problems.

- *'I made no mistakes, and I got an OBE for my efforts.'*

That's very nice! It *is* possible to beat the system, after all. It's good to find someone who has been recognized. All too often the gong is posthumous; it doesn't look as good on a gravestone as on a living achiever's chest.

Try Another Way

The inventor has to be more than an innovator. Looking at the society and commercial market in which the invention needs placing is part of the game. If the idea doesn't fit the slot for which it was originally conceived, maybe it can do well somewhere else. It's also possible that 'somewhere else' will turn out to be more profitable. One change of angle is to ask if the invention is being sold to the right manufacturer.

There's an improvement to portable car ramps which stops them moving away under the pressure created as the car drives on to them – a well-known problem. Strips of webbing are attached to the far end of the ramp's wheel-tracks. This webbing runs along the tops of the ramp's twin surfaces and forward over the ground in front of them. As the car moves towards the ramp, its weight – now imposed on the webbing – holds the ramp in position so that it cannot slide away.

It's not a bad idea, but the makers of ramps weren't interested. Why should they be, when they sell successfully without it? An alternative approach for the inventor could be to tackle companies which make webbing, as the idea offers a new outlet for their material.

Another inventor designed a new tilt head to work with a painters' easel. It allows the canvas to be adjusted to any position the artist finds comfortable. Even if the easel's feet are placed on uneven ground, making it stand at an awkward angle, the tilt head straightens the canvas up. It relies on a balanced screwing device which, sadly, turns out to have no buyers in the art field.

Again, the innovator should look for a new manufacturer where the tilt head might sell well. Professional photographers in the fields of film, video and stills need the best tilt heads they can find, and this one would be ideal for them.

The most striking example of change of use I know is the cat flap cited on page 8. Anyone who owns a cat – and a flap – knows how the flap bangs noisily around, and how it lets cold air rush through the house whenever the wind is in the wrong direction.

Normal flaps turn on an upper pivot so that the cat can push the flap to enter or leave the house. The new flap turns *sideways* on a central swivel. The cat pushes one side to depart outwards and the other to return inwards. The aerodynamics of the design prevent it banging in the breeze. I was so impressed by this idea that I installed one at home for my own moggie, and it has worked perfectly for many years. But it achieved no commercial success.

Then some genius saw that it could be used in an entirely different way. A king-size version is now installed at a hospital in Reading, where it works as a swing door. Hospitals have the same problem as cat-flap owners, but on a much larger scale. They have big doors that let in a lot of undesirable noise and cold air. The man-sized cat flap now solves those problems, letting stretchers and trolleys move in and out.

• *'I was guilty of not looking at unusual alternatives e.g. I ignored an offer from a washing machine gearbox manufacturer to look at my vehicle servo-synchromesh, even when all others had failed to take an interest.'*

An invention designed in one direction is often much better working in another place. Inventions can have breadth as well as depth. *Mea culpa*. Look at my own stupidity with the hydraulic control valve which ended up being applied to automatic lathes (see page 78) to the profit of other people than myself. The difference in the case of the inventor who writes here is that he actually had an alternative offer. Never turn down an offer without evaluating it extremely carefully.

CHAPTER THIRTEEN

Should We Change the System?

What is the future of inventing in the United Kingdom? The country's record is glorious, as is confirmed by the report from the Japanese Ministry of International Trade and Industry which opens this book. More than half the successful inventions since the Second World War came originally from the UK.

There are suggestions that the British are losing their grip. The United Nations Educational, Scientific, and Cultural Organization (UNESCO) reports that the UK has one of the poorest current track records for filing patents of any advanced industrial country.

UNESCO's league table shows that Japan files forty-three patents per 100,000 population, followed by Switzerland with forty. Then come Germany and Sweden, the USA, Austria, Finland and France. The UK belongs in a lowly group, including Norway and Denmark, the Netherlands, Belgium, Greece and Spain, which files between four and eight patents per 100,000 people i.e. at best, a fifth as many as the Japanese and Swiss. UNESCO says that only Ireland, Italy, Turkey and Portugal do worse. The UK's share of US patents has dropped by nearly half in the past twenty years.

And yet, despite their poor showing with patents, UK scientists and engineers publish more papers on their work than any other European country. Behind the UK in descending order are Germany, France, Italy, the Netherlands, Sweden and Switzerland. This suggests that British inventiveness is *not* dead. Perhaps the UK's shortage of written guides to taking a bright idea through patenting to successful exploitation and sale is one cause of the problem. This book is designed to help.

NON-COMMERCIAL AID

Could the system itself be changed or improved in some way? Over half the innovators who answered my survey want some kind of independent body or panel which evaluates inventions and helps in their development.

• *'We need a professional body where inventors could possibly pay £50–£100 to get a confidential assessment of their ideas for marketability and potential before making any major investment. Those ideas would receive a written assessment of the pros and cons, so the novice innovator would have some knowledge of what he was letting himself in for.'*

Expert assessors cost money, quite a lot of it – £100 would get you nowhere. You certainly won't find the kind of service you are looking for, with evaluation, comparison and written reports, at a cost below what an accountant would demand for a day's work. Members of the Institute of Patentees and Inventors have a degree of assistance on offer to help assess ideas (although there is no formal panel of the type envisaged by the writer).

• *'I feel that the DTI could – under its auspices – form vetting committees who, should they feel an invention submitted to them was of particular merit, could both sanction reduced patent fees, provide professional help and assistance in development of the idea. I would also require large PLCs to have appointed a person whose responsibility it would be to view new innovations submitted and give fair assessment of their worth to that particular industry or profession.'*

The government (which includes the Department of Trade and Industry) is bowing out of playing Auntie. Its Business Link scheme should be of some help to innovatory businessmen but direct aid to inventors is becoming almost non-existent.

There have been government-backed organizations which were set up especially to help the struggling inventor's idea to reach the market – NRDC and BTG, for example. They ended up behaving like any private business, creaming off the very good ideas and sending all others away with a warm letter. The most warmth you could get from the letter was by burning it.

• *'What is required is a sponsorship publication. There would be a list of companies/organizations etc prepared to examine an innovation/invention/idea in detail and give a full and comprehensive reply to an applicant, as far as the information supplied will allow.*

'Such a company, etc would possibly back the inventor either financially and/or technically, or possibly recommend or put the applicant in touch with a suitable backer, or alternatively advise what action should be taken, including abandonment.

'Applications could be channelled through an appropriate body which could be set up as a department of the Institute of Patentees and Inventors, which would also monitor the quality, progress etc of the examinations being made by a particular company.'

An invention can be as simple as the perforations which make postage stamps separate; it can be as complex as a swing-wing aircraft. Who are the experts who can cover every possibility that may come up? Where do you get them together? How are they paid for?

A publication implies the existence of thousands of companies who are willing to assess innovations. There are certainly some, but not enough. Past attempts to establish a mass database have failed.

The suggestion for 'an appropriate body . . . which would monitor the quality, progress etc of the examinations' implies a whole Institute, charged to the tax payer, devoted to the study of inventions. The mechanics of funding it would require the activities of a brigade of administrators, also paid for by the tax payer.

The proper test of an invention is the market-place. From society's point of view, it is also the most efficient means of testing.

Change The Patent System

• *'I have the strongest possible views on the national treatment of inventors. Inventiveness is a quality possessed as a gift and possessed by very few but liable to enhance the national benefit. The inventor should therefore not be called upon to pay anything by way of professional cover and other fees.'*

'Inventiveness is a quality possessed as a gift.' No. It's something all of us have, children have it ad lib. The trouble is that many of us lose it when we grow up. However, the suggestion that patenting should be made free *is* practical and easy to administer, unlike most of the ideas offered for improvements. It would certainly increase the number of patents filed per head of population.

But who will pay for it? The tax payers? Even if they agree, will tax payers be happy about giving free patents to giant industrial corporations and to foreign companies? I don't think so. That means there has to be some mechanism for distinguishing between deserving private inventors and undeserving big companies. Here comes the bureaucracy again, and already the costs are beginning to rise.

Then there's the little problem of establishing which citizenship qualifies patent applicants for the freebie. UK, you say? But what about Europe? Won't the European Union have something to say about unfair competition? In practical terms, I fear the idea is a non-runner.

- *'Allow inventors to pay renewal fees in the form of equity in the invention.'*

That's asking the people who give the patent to become gamblers on the success or failure of the innovation, to become interested parties.

- *'Let's have a quicker patenting process.'*

It's almost impossible to arrange 'a quicker patenting process'.

The whole business starts with research in the Science Reference and Information Service library. That takes time to carry out properly because, if the work is skimped, you can end up re-inventing something which already exists. The year's wait after submitting a patent application is there to allow time for testing the market, finding a manufacturer, raising capital and organizing everything needed to make an invention successful. After that, the Patent Office itself requires time to make its own assessment of the application. This work cannot greatly be speeded up by machines; it is an art as much as a science.

And yet changes *are* taking place on a small scale. Computerization of databases and the advent of CD-ROM is making it simpler and swifter to call up records for examination. The European Patent Office reckons to issue a patent in just over three years.

Publicity

- *'For one year, all new inventions by small companies or individuals should have free publicity to the public.'*

Who's going to give the free publicity? The newspapers? Television and radio? Rupert Murdoch? It's not their job to do that. If you invent a bridge that falls down, you'll get all the free publicity you can handle, and some more. But good news is no news unless it is world-shattering.

Or is the idea that the state pays for free advertising? Does this mean that the state also pays advertising agencies to design the adverts and 'place' them in various publications? Could paid 'consultants' be required to advise small companies and individuals on how their 'free' advertising should be organized to best effect? This could be one of the biggest gravy trains in history! Can I have a ticket, please? Free, of course.

- *'TV programmes might encourage new practical inventions.'*

Yes. Give me some more TV programmes about inventions and I'll be happy to present them! One of the programmes I made on inventing is now used as a piece of course work by the Open University. Television is a field where I am too closely involved to give you an objective answer, but from the TV companies' point of view the crucial question is 'How many viewers will it get?'

I suspect that giving inventing a higher profile, looking at it from a constructive point of view, especially demonstrating the achievements of younger people, would do nothing but good, and would attract large, devoted audiences just as the BBC's *Young Scientists of the Year* series did twenty years ago.

BETTER EDUCATION

- *'Encourage the concept of innovation in schools – demonstrate the road from simple concept through to commercial success. Reward useful and innovative ideas at an early stage – invention/innovation could form part of 'technology' which subject now forms part of the core curriculum (my wife teaches this new subject!).'*

It's happening. Innovation is becoming understood as part of an economic process without which we will all be much poorer. The technology courses cited above are an important part of this and it's most encouraging to see the way in which they are turning out young people with good ideas – especially the girls.

There's also a new tendency for industry to go to colleges to obtain fresh insights into its own problems. One example I came across was a manufacturer who had a difficulty with moving frozen meat on conveyor belts. It sticks. The young people suggested that he should get rid of his stainless-steel equipment and replace it with wood, specially treated so that it can be steam-cleaned.

Brilliant! Wood has a lower thermal conductivity, so sticking is reduced. Nobody long established in the freezer business would have come up with the suggestion because they are too conditioned by their experience of using stainless steel as standard.

'Chance Would Be A Fine Thing' Department

• *'Shake up the City to provide longer-term risk money.'*

You're joking, of course!

New Kinds Of Patent

• *'If your idea is not presently exploited commercially you should be granted sole rights of exploitation for a period of say seven to ten years.'*

This is what is known as an Innovation Warrant. With a full patent, the idea must be new and not obvious. A warrant would require only that the product is not already on sale – the point about obviousness would not arise. The warrant could not be challenged once it had been granted. A suggested cherry on the cake is that the warrant would be *protected by the state*, unlike a patent, whose defence is the responsibility of the patent's owner.

If this idea is to be any more than minor tinkering inside a single nation's patent system, it involves a change which is almost impossible to implement. Worldwide agreement is needed to add a new factor to a patent system which has evolved slowly over centuries. Without a new international Patents Convention Treaty, the proposal is dead in the water.

Another possibility is the so-called Petty Patent or registered invention. This would give a smaller degree of protection than a full patent, but at a lower cost and for a shorter time. Anything awarded a petty patent would be less novel and be more obvious than a patentable idea.

A hypothetical example might be a new treatment for car steering wheels. On hot days, wheels tend to slip in sweat-soaked hands, lessening the driver's control over the vehicle's movements; but the wheel still needs to slip freely through the fingers as it returns to its normal position after turning a corner. An innovator meets the problem by slightly roughening the upper surface of the wheel while leaving the underside smooth as before. Assuming that it works, the idea is not worth a full patent but might be worth a lesser amount of protection.

The petty patent sounded a plausible idea when it was being actively canvassed in the 1980s. Its essence was incorporated into the Copyright, Designs and Patents Act of 1988 and the roughened wheel could now be protected as a Registered Design.

One attractive thought has nothing to do with changes in the law and everything to do with making human nature work to the inventor's benefit. It is plausible because it goes with the system instead of fighting it. Football pools companies would be asked to run a lottery based on invention. They would pay the DTI or the IPI to make a list of fifty projects – checked and authenticated as genuine. Punters would back the project of their choice and their betting money be applied *pro rata* to the projects. Successful projects would pay off to the 'investors'. This way, even losers would get the feeling that they are doing good with their flutter, the winners win, no one loses their shirt and the country profits.

The ideas discussed in this chapter are mostly non-starters and, at best, palliatives, though some of them are better than nothing. None makes a massive difference to the major problem, which is that too many innovators fail to find success. The hard truth, and many inventors will find it *very* hard to digest, is that they themselves are at fault. They are good at finding new ideas but bad at realizing them. They do not understand the manifold problems of converting an idea into a product that sells and makes a profit in the market-place. The most practical way to improve the system is to improve the capability of the people who work in it. If they become professional, the playing field becomes more even, and difficulties easier to handle. *We don't need a better deal with patents; we need better informed innovators.* Comments from two members of the IPI who have steered their way between the Scylla of incompetence and the Charybdis of unrealistic hope to find the harbour of financial success, back up this judgement:

- *'Having been an entrepreneur and innovator for over thirty years, I feel that the difficulties of initial financing and launch problems can generally be overcome, or by-passed, if one is determined enough and if it can be shown that the idea or product has unique merit.'*

Explore the system, come to terms with it and work it. It's the only practical way to operate, and much easier than complaining. The existing world is the world we've got. It is almost impossible to change, so learn to use it to best advantage.

- *'Inventors should be (a) patent-wise (b) presentation-wise (c) reality-wise.'*

Ron Hickman – for it is he – says it all. I've been trying to get this message over for many years. We should be dreamers, but not only dreamers. Learn what you find useful from this book – and keep trying!

Rule Ten: Persevere

Is the message never to despair or is there a point at which to give up? The answer is: both.

There comes a time when you must be ready to throw away your *idée fixe*. Do it without regret. But how do you know when the point is reached? One way is to estimate what money the product is likely to earn over the next few years, and how much you can afford to lose if it fails. In concrete terms, say to yourself: 'I cost my own time at £X an hour. I will/will not invest more.'

A good time for self-questioning is before the twelve-month period runs out on the provisional patent. If, in a year, you have been unable to attract any meaningful outside interest in your invention, could it be that it contains some commercial flaw? If you think so, withdraw your patent application. Then no outsider will ever see it. That gives time to go back to the drawing-board and make radical changes to the invention; alternatively, the idea can be junked at no further expense.

But *perseverance* is the quality which creates winners. Persist if there is a demand for the product, and if it can be made at the right price.

Case history

The story of Terry Payne and Monodraught encapsulates so many of the problems facing people struggling to introduce a new product that it is worth telling at length. Terry Payne was actually a licensee, not the original inventor, but he took the product over after its inventor's death and has made it his own. He tells the story himself with the addition of a few comments from me as he goes along. The key word is *perseverance*.

'Bill Stranks was an aeronautical engineer who lived in Norwich and he realized there must be a reason why some chimneys smoke and others don't, although they may be identical in height or overall design.

'He knew that it was all to do with the differential wind pressure between the top of a chimney and the fireplace and he, therefore, initially invented a system which required a large "balancing chamber" behind the fireplace to equal out this differential air pressure.

'Subsequently, he realized that it would be very expensive to supply this balancing chamber for private houses but that, if he applied his same principle to a closed boiler house, he could achieve the same beneficial effect.'

Comment: Monodraught eliminates the need for ugly, tall chimneys conveying exhaust gases from commercial boiler houses. In the past, it was necessary to put the outlet above the highest part of the building (the roof level) to avoid problems with downdraught. Bill Stranks *identified a problem* and *met a need*.

Monodraught needs no more than a metre of chimney height and can be positioned on any single-storey building, regardless of the presence of higher buildings nearby (see photo section). It takes air through slotted louvres at the top of the chimney and passes it down through the outer cavity in the chimney casing, to the boiler room below. Exhaust gases rise through the inner tube of the twin-wall flue and are vented to atmosphere.

No matter which way the wind is blowing, or how much its direction shifts, Monodraught's circular arrangement of louvres ensures that pressure changes are compensated automatically. The proper operation of the system remains unaffected.

'In 1973 I had a small building business in High Wycombe and a consulting engineer instructed us to buy one of these Stranks Monodraught systems for a boiler house which was sited at the base of a chalk pit.

'We installed the Monodraught but the boiler manufacturers refused to commission the boilers since they said this little chimney couldn't possibly work and that we would have to take the chimney to the top of this chalk pit.'

Comment: Not Invented Here!

'The client insisted however that the Monodraught would work and when we forced the boiler manufacturers to come back (with their threat that they would charge us for all their abortive time) the boilers were then commissioned and the manufacturers said they had never known the boilers operate at such a high efficiency! They insisted however that it had nothing to do with the Monodraught system!'

Comment: NIH sets people's attitudes in century-old concrete. Nothing can blast them loose from their opinions.

'We did two more jobs in quick succession on "impossible situations" but on the fourth occasion Mrs Stranks told us that Bill Stranks had died and she was trying to wind the business up. I thought it was a tragedy that such a brilliant idea should just be buried and forgotten, so we arranged to manufacture the system under licence.'

Comment: Monodraught found its product champion though,

sadly, not until after its inventor's death.

'In the early days we certainly did have an uphill struggle, trying to convince people that the system would actually do what we claimed. Everyone's perception of a "chimney" is something that sits on the roof of the very highest part of the building. To convince people that this concept was no longer necessary was very difficult.

'It was on 18th September 1975 that the Monodraught appeared on *Tomorrow's World*. This was without doubt the making of Monodraught since it gave it the credibility it had sought for so long. It was amazing how many years afterwards, at exhibitions and meetings, people would say, "Yes, I remember seeing this on *Tomorrow's World*"! One of my treasured possessions is a copy of the old script.'

Comment: I can't say how much pleasure these comments give me and my co-author! I was responsible for writing the Monodraught script; Robin Bootle directed the live outside broadcast.

Terry Payne was a trifle lucky because the programme, in my own view, was one of the best episodes of *Tomorrow's World* ever made. But you make your own luck sometimes. If it's going your way, ride it. Advertising in the right place pays off.

A problem that now arose was the fact that Monodraught was so different from existing flues that existing regulations and standards did not cover it.

'I got elected on to some of the committees of the Code Drafting Panels of the British Standards Institution to put my case more forcibly. They subsequently appointed a working party and at this stage I just kept plying them with the number of contracts which had been carried out and I said: "Either approve it or ban it, but please don't sit on the fence any longer!" They accordingly inspected over sixty working examples of Monodraught systems, none of which they could find malfunctioning in any way, despite theoretically breaking all the golden rules. Accordingly, we had seventeen new clauses written into the British Standard specifically to cover the Monodraught.'

Comment: An outstanding piece of work! To persuade the authorities to rewrite the rule book in your own favour is given to few of us. Of course, it wasn't *given*; Terry Payne worked hard for it.

'For the first ten years we simply made concrete block systems.

Because of the size of the larger systems, weight was always a problem and our lightweight glass-fibre systems were launched in 1985. These have proved a spectacular success.

'Bill Stranks' patent expired in 1981 and since that time we have relied on Registered Design protection, on the basis that we now have so many variations on a theme that it is Monodraught's appearance which is more easy to defend.'

Comment: This is exactly what I have advocated throughout this book. Keep improving the product, then it doesn't matter so much if the patent runs out. You'll have a strong market position of your own by that time. And there are more ways than one to kill a cat. Monodraught intelligently switched to the use of Registered Design protection at the right moment.

'Someone telephoned me and said: "I am not sure whether I've got the right firm or not, but do you make something that looks like a dustbin that sits on the roof and takes in fresh air?" '

'I knew at that stage that we would have to do something about the design of our systems!'

'I was sitting in a park in High Wycombe eating a sandwich and I looked up at the Town Hall and saw a louvred turret in a classical shape. It made me realize that we were only one step away from combining all the advantages of the original Monodraught principle with a range of decorative turret designs that could become the focal point of a building.' (See photo section.)

Comment: You see how Terry Payne is forever driving forward, changing plan to meet changing circumstance, always on the attack, rather than defending an existing position. Under his guidance Monodraught has become a highly successful company. And he has the satisfaction of working with a basic commodity which is generally regarded as totally unattractive, and making it so sympathetic that it can become the attractive focal point of the building. There are even designs which are used on many listed buildings.

'Perhaps the greatest kick I have got out of the development of the company over the last twenty years is the wonderful phrase that people say: "But it is so simple, why didn't somebody else think of it before?" '

'We have had a lot of fun out of the development of the system. It has been exciting to swim against the tide and pioneer something which has such beneficial effects.'

Comment: And if you're not having fun, why are you in business?

Terry Payne makes a final, telling remark: 'I have had my fill of bank managers and accountants who tried to persuade me in the early days to give up "because it was not a viable proposition".'

Comment: All together now. Repeat after me: Rule Ten: Persevere.

CHAPTER FOURTEEN

Is the Theory Right?

So far I've shown you how to train as an innovator, how other innovators succeed in a cold, hard world, how to present ideas and how to choose the most suitable protection for them from a tangle of patent, copyright, and design rights. We've talked about *you*. I've tried to help you plumb your own feelings – what you really want to gain from being an innovator – and to prevent you engaging in unnecessary fights. I hope you now have the right ideas about approaching manufacturers and holding their attention.

Everything I have said is based on a lifetime of experience. My respected colleagues in the inventing business will have arguments to pick with some of the points I make, but I know that they will be in broad agreement with my message, because it covers problems which have kept us debating into the small hours on many occasions.

However, that is no reason to accept my ideas on trust. One of my messages is 'Never take anything for granted'. Check everything, especially if it is written down and looks authoritative, is my cry. This means that I must be ready to follow my own advice and try to offer firm proof that what I say is sound.

The first part of this book is concerned with innovating and inventing today. Now let's discover what lessons we can learn from the successes and failures of inventors in the past. What rules will emerge to typify their struggles and experiences? Will they match the Ten Rules I have suggested for modern inventors?

Avoid the common trap of assuming that, because an event happened long ago, there is nothing to learn from it. The process and principles of innovation do not change with the passage of time; past inventors were as bright, determined, quirky and, above all, as human, as us.

Suppose that a time machine could carry you on a visit to your great-great-great-great-grandchildren. You would rightly be irritated to discover they regarded you as some variety of freak, simply because you lived two hundred years before them. Avoid offering the same insult to your ancestors. *They* were the people who devised inventions which make life comfortable in industrialized countries today. Think of the wonderful legacy they left us:

- Electricity. Lights, saving us from having to switch off personally when the sun goes down and adding many active hours to the day. Power, driving essential machines in factories.
- Gas. First made from coal, now supplied from wells beneath the sea, cooking our food and heating the buildings where we work and live.
- Cars and trains, carrying us beyond the horse-ride limit which condemned earlier generations to lives within thirty miles of home.
- Aeroplanes, enabling us to travel to the far side of the planet with hardly a thought for the distances involved.
- Telephones, computers and faxes, bringing the world into our workplaces and living rooms.
- Fresh water from taps, and mass sewerage systems, creating high standards of public health in cities formerly prey to epidemics of cholera and plague.

What can we learn from the past to help inventors today? With the benefit of hindsight, can we establish any rules from what happened to earlier innovators and their ideas? Here is the supreme test of my theories. I can't shirk its challenge. It is time to test my second set of Ten Rules.

Bob Symes' Second Ten Rules for Inventors

Rule One
Invent for tomorrow

Rule Two
Don't be before your time

Rule Three
Experience is never wasted

Rule Four
Join a network

Rule Five
Prove your invention works

Rule Six
Don't become a dead hero

Rule Seven
Operate the system

Rule Eight
Get a manufacturer's backing

Rule Nine
Success depends on the inventor

Rule Ten
Persistence leads to success

Invention Creates History

The idea that invention creates change, and that change can be expected to alter the world in every generation, is comparatively new, less than 300 years old. In past societies, where nine out of ten people were engaged in farming and food-getting, the day was measured by the position of the sun in the sky, and the slow turning of the year was marked by the progress of the seasons. Crops were gathered in, taxes paid to the lord or king. Nothing altered. Sons lived the same life as their fathers and grandfathers, wives performed the same tasks in the same way as their mothers-in-law.

Even among those fortunate people with wealth and leisure to indulge in philosophy and debate, human achievement was measured in terms of the biggest or best, rather than in terms of doing the greatest good or the most useful achievement or valuable new discovery. The idea of novelty, using the word in its proper sense of 'newness', did not enter commentators' minds.

There's a lot to be learned from what provoked wonder in the Ancient Greeks, the most intelligent and inquisitive European people in the centuries before Christ.

The Seven Wonders of the World

The earliest known list, probably put together from guidebooks, which, believe it or not, were popular reading in Greek society, was compiled by a writer called Antipater of Sidon. He based it on the world as known to the people of the Greek city of Alexandria in Egypt in the second century BC. This probably explains why two of the Seven Wonders are in Egypt and five are of Greek origin!

The Tomb of Mausolos

Mausolos was king of Caria in Asia Minor – nowadays western Turkey, bordering the Mediterranean. After his death in 353 BC, his widow Artemisia erected a tomb to his memory at Halicarnassos so magnificant that it became one of Antipater's Wonders and later gave its name to the English language: 'mausoleum'.

Don't think of the tomb as an ordinary monument. It had architects – two gentlemen named Satyrus and Pythis – and was a building in its own right, with a tremendous statue of Mausolos in his chariot on the roof. The Mausoleum collapsed in an earthquake before the fifteenth century, and sculptures and reliefs from it are now in the British Museum.

The Lighthouse (Pharos) of Alexandria

Alexandria was founded by and named after Alexander the Great. Construction of the lighthouse began under his successor, the general Ptolemy, who became Greek pharaoh of Egypt, and was completed by Ptolemy's son around 280 BC.

The lighthouse, which was the prototype for all others built since then, stood on the island of Pharos, nearly a mile from the coast of Africa. Estimates of its height vary between 300 and 450 feet. At its top, a light burned which was said to be visible nearly thirty miles out to sea. It guided ships towards the harbour of Alexandria rather than, as do modern lighthouses, warning them to stay clear of danger.

The lighthouse (or Pharos, as it was known) survived into the second millennium AD, meeting its end in an earthquake in 1375. Before then, Idrisi, an Arab geographer, noted that its blocks of stone were held together with lead ties – a method reintroduced during the building of the Eddystone lighthouse, 2000 years after the Pharos was erected!

The Colossus of Rhodes

A metal statue of Helios, the Greek sun god, 105 feet high, the Colossus was designed by the sculptor Chares of Lindos (a city of Rhodes) and completed in 280 BC in gratitude to Helios for saving the city from a siege. It was cast from the bronze weapons and equipment abandoned by the besiegers.

The Colossus acted as a landmark to sailors, but only until 224 BC, when it was felled by an earthquake. Its toppled remains continued to excite wonder until AD 656, when Rhodes was occupied by the Saracens. They sold the scrap to a merchant, who needed 900 camels to carry it away. The statue did *not* bestride the harbour with ships sailing between its legs, as has often been depicted. That was a feat beyond the capability of ancient engineering.

The Temple of Diana at Ephesus

Like the Mausoleum, the Temple of Diana was in western Asia Minor, which was culturally part of Greece until it was occupied by the Turks. It finished before 300 BC and belonged to a succession of buildings dedicated to Diana or Artemis, one of the most popular goddesses of the ancient world.

The roof was supported by stone columns sixty feet high and the interior was ornamented with works of art by the greatest of Greek sculptors, Phidias and Praxiteles. St Paul knew the temple. He spent a couple of years living in Ephesus, preaching Christianity, and was driven out by a mob hired by local silversmiths, whose living from selling figures of the goddess was threatened by his activities. 'Great is Diana of the Ephesians!' was their cry.

The building was destroyed by the Goths in AD 262 but some of its columns are in the British Museum.

The Statue of Zeus at Olympia

The only Wonder from mainland Greece itself. The forty-foot-high statue was made by the same Phidias whose work adorned the Temple of Diana at Ephesus. The god was seated on a throne, holding a sceptre. His body was made of ivory, his robe of gold, and the statue reached almost to the ceiling of its temple. No wonder all trace of it has vanished; it must have been irresistible to plunderers.

The Hanging Gardens of Babylon

Built by Nebuchadnezzar, King of Babylon, around 600 BC, the Hanging Gardens were a series of square terraces each measuring 400 Greek feet to a side and each stepped back from the one below. They were planted with trees and flowers from all parts of the Babylonian empire whose luxuriant foliage hid the terraces and so seemed to hang in the air.

The story is that Nebuchadnezzar built the Hanging Gardens to please one of his wives but as he made Babylon the queen of nations, rebuilding the city and restoring temples throughout the land, it seems likely that they were simply part of his public works programme.

Modern excavations of the city ruins revealed a palace with irrigation works which may have been the site of the Gardens.

The Pyramids

The three Pyramids at Gizeh, in Egypt, are the only ancient Wonder still in existence, although they are far and away the oldest, dating from between 2600 and 2500 BC. The Great Pyramid, that of the pharaoh Khufu, was the first; the others were built by his son and grandson, Khafra and Menkaure.

The Egyptians believed in a physical afterlife where their resurrected bodies required food, drink and all the luxuries which make death worthwhile. The purpose of a pyramid was to protect the royal occupant's body after death, along with all his possessions.

The more you contemplate the Great Pyramid, the more mind-boggling its statistics become. It covers an area of thirteen acres; its stone weighs nearly six million tons; its cubic contents measure three million cubic yards. Before it was stripped of its gleaming casing of white limestone, the length of the sides of its base was 755 feet (more than an eighth of a mile) and its height was 479 feet. Today it is 450 feet – fifteen times as high as the average house.

The outside is composed of limestone and granite; this layer contains more than two million blocks of stone, each of which averages two and a half tons in weight. If the stones were cut into blocks one foot square, laid end to end, they would run two-thirds of the way round the earth at the equator.

The Seven Wonders had a number of features in common:

- They were all buildings or statues (three of which were destroyed in earthquakes).

- Five had religious significance (the exceptions being the Hanging Gardens and the Pharos). Two were tombs, two were statues erected for reasons of worship or to offer thanks to a god, and one was a temple.

- The driving personalities (kings, queens, pharaohs – despots all, apart from Chares of Lindos) behind the construction of five of them are known to us by name, though they all lived over 2000 years ago and Khufu of the Pyramids twice as long ago. The Wonders were a triumph of the personality cult.

- Most of the Wonders have disappeared, but they were long-lived. The Hanging Gardens lasted 300 years, the Temple of Diana over 500 and the Colossus of Rhodes, despite being overthrown within sixty years of its dedication, could still be inspected in ruined form 800 years later. The Pharos and the Mausoleum both survived their 1500th anniversaries and the Pyramids are with us yet.

Those are the more striking of the positive aspects shared by the Seven Wonders. Their negative features are equally marked.

- None was designed for the public good or use – except the Pharos of Alexandria, a beacon for navigators. The Colossus of Rhodes acted as a landmark, but that was not its purpose; it commemorated and expressed thanks for the favourable intervention of the god Helios in the siege of Rhodes.

- There was no novelty involved in six of the Wonders' construction, although in some cases known techniques were pushed farther than before. The Pyramids, for instance, came at the end of a long process of design evolution. The odd one out was the Pharos. Apart from the fact that it was the ancient equivalent of a skyscraper and its binding with lead ties was only paralleled 2000 years later, it was the start of a family of lighthouses based on its design, rather than being the dying culmination of an old tradition.

• None of the Wonders used energy, apart from the Hanging Gardens – which must have had some lifting mechanism to carry water to the top for irrigation purposes – and the Pharos at Alexandria, with its beacon burning on the uppermost platform. One intriguing possibility is that its architect, Sostratus of Cnidos, designed a mirror system to concentrate the beams of light given out by the fire and control the direction in which they pointed out to sea. Without some such device, it is difficult to credit the suggestion that the light could be seen thirty miles away at night.

• None of the Seven Wonders – apart from the Hanging Gardens with its irrigation system and, speculatively, the Pharos with its mirror – had a single moving part, or any piece of engineering or design which fitted into and worked with others.

A clear picture emerges from the list. Here was no concept of change, no idea of improvement, no thought that there might be improved ways of living. Even the real technical achievements of the ancients did not enter the mind of the list's compiler.

Amended versions of the Seven Wonders were no better. As time went by and geographical knowledge spread, the Colosseum in Rome and the Great Wall of China appeared on the roster but inventions did not. There *were* inventions, but society was incapable of seeing them as wonderful.

The only Wonder with any claim to be an agent of betterment or change was the Pharos. It also seems to have had various technical successes to its credit. But no one noticed that it differed in quality from its six colleagues.

The Seven Wonders never included the Romans' continent-spanning highways or their superbly engineered water-supply systems, although tourists today travel for thousands of miles to view the Appian Way or the Pont du Gard. Never listed was the amazing eightfold water-wheel system on the hillside at Barbegal in southern France – which ground corn to supply the requirements of Roman Arles.

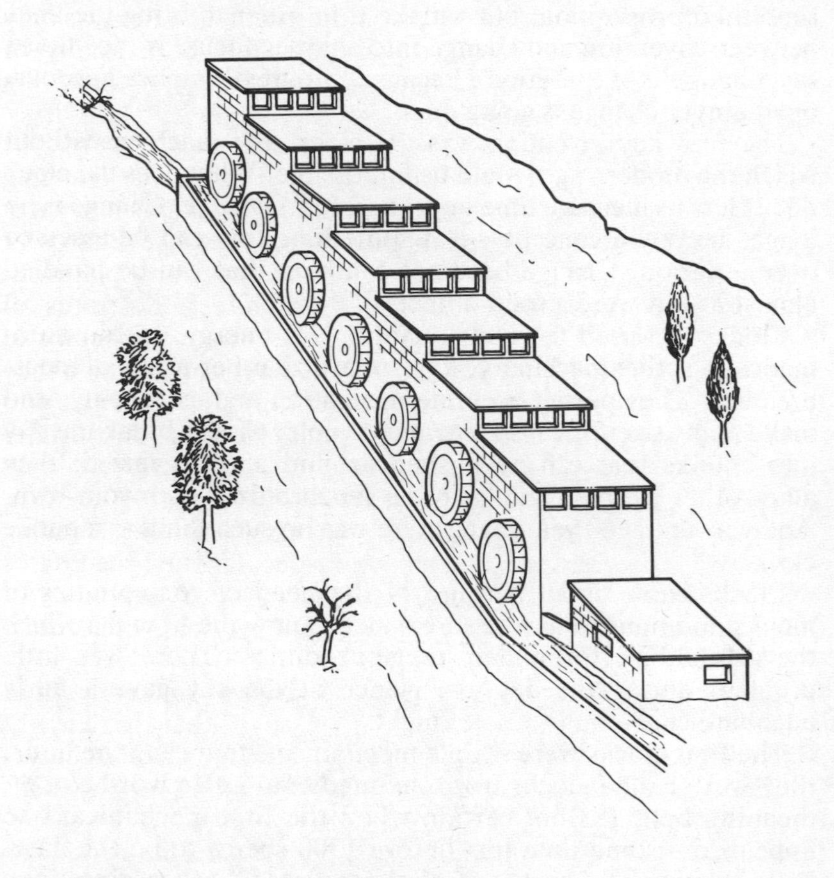

Not mentioned was the automatic grain-harvesting machine shown carved on a stone at Buzenol in Belgium and described thus by the fifth-century writer Palladius: 'With the aid of a single ox, the machine outdoes the work of labourers and shortens the time of the whole harvesting operation.'

Although invention did exist and happened at a steadily accelerating pace from AD 1000 onwards, it was not until the eighteenth century that its world-changing possibilities were fully recognized. Had the matter been otherwise, the catalogue of Seven Wonders would soon have been adjusted. The fact that it was never altered shows that the thought pattern which created it did not change either.

THE CLOCK

Two inventions, several hundred years apart, were crucial to the transformation of ancient to modern thinking. A grossly over-simplified proposition, but I make it in order to bring the links between invention and change into sharper focus. Notice that I say 'change' not 'progress', because 'progress' implies approval of whatever change is going on.

The first key invention was the clock, the machine without which the modern age would be impossible. What does the clock do? It lets us measure time and, therefore, *change*. Change is no longer merely a concept – it is something that can be assessed over a period. Change becomes a notion that can be handled almost as if it were a solid object.

Clocks are used to count quantities of energy, to run automatically acting machinery, to standardize other forms of measurement. They permit accurate navigation and astronomy, and make many scientific instruments possible. Clocks break the day into chunks that can be pushed around and organized; they allow other people's chunks to be synchronized with your own. And yet, until 700 years ago, there was no such thing as a public clock.

Clocks came about because of the need of communities of monks and nuns to have a sure way to know the hours at which they should perform their religious duties. There was little problem during the day – a glance at the sky gave a fairly adequate clue – but not so at night.

The first clocks were simple mechanisms that rang the hour; they were called clocks from the medieval Latin word *clocca*, meaning bell. It's not certain when the first mechanical one appeared – some time just before 1300 seems to be the date. It took time to sort out the process of measuring time properly. Originally, both days and nights were divided into twelve hours. This meant that a night-time hour in June was a good deal shorter than a day-time hour in June; equally, a night-time hour in June was shorter than a night-time hour in December. Once clocks were invented, this way madness lay. But then, in 1335, the first clock recorded as striking *equal* hours by day and night appeared in a church in Milan.

Inventing the clock would not have been sufficient to cause change, if there had not also been a new willingness to accept the change. The church itself did so and town dwellers showed the same willingness. They began to rise at a known time, to open their shops during fixed hours, to make business appointments for when the clock struck three.

The chiming of medieval church clocks started western

civilization on the long road towards today's world of invention, where time can be known in millionths of a second or as a quantity which itself changes at speeds approaching those of light. Measurable change, if not always welcome, is expected as a matter of routine.

The Beam Engine

The second vital invention was that of the beam engine by Thomas Newcomen in 1712. Later work by James Watt on improving Newcomen's engine has tended to obscure its importance in the eyes of later generations, including our own. But it is not too strong a statement to say that the arrival of Newcomen's engine marks the beginning of the process of steadily accelerating invention that makes life so exciting and sometimes stressful today.

The invention itself was extraordinary. For the first time it became possible to use power not only in much greater quantities than any living creature could deliver, but power that never tired. Moreover, it could be generated anywhere it was needed. No more were men dependent on the whims of the wind, the variable flow of rivers, or the limitations of domesticated animals.

Overnight the world changed, and the beam engine was not only the cause of that change, but became an active agent promoting it. It was used to pump out flooded coalmines. This produced more coal which, converted to coke, allowed greater quantities of iron to be smelted. (Can it be coincidence that Abraham Darby's coke-fired blast furnace also saw the light of day in 1712?)

Cheaper and better iron in turn made it possible to produce machines – including improved beam engines – which carried the Industrial Revolution forward. Larger quantities of available coal led to William Murdoch's use of coal gas for lighting. And so the process of change accelerated.

By the time James Watt died, a century after Newcomen's invention, society recognized what was happening – so much so that energetic efforts were made to preserve Watt's scientific papers as a memorial to his genius. It was a process that, ironically, led to the partial burial of Newcomen's reputation.

'Tomorrow' began to exist as an active idea meaning a day reflecting change from today; inventors, with their ability to identify a need, became its creators. This brings us to the first of my second set of Ten Rules for Inventors:

Rule One: Invent for tomorrow

If we catalogue today's Seven Wonders of the World, how do they compare with the originals? We will each have candidates for inclusion in the list, depending on who we are and where we come from, but some points are sure:

- Few, if any, will be buildings and none will be statues.

- Churches, temples or tombs no longer qualify in an increasingly materialistic world.

- Nothing which displays conspicuous consumption or self-promotion by super-leaders (communists or capitalists), or objects designed for private use or pleasure, will be included.

My list is as follows:

- Communications satellites
- Computers
- Body scanners
- Television
- Lasers
- Genetic engineering
- Applied nuclear energy

If you feel otherwise, I won't argue. For instance, if you have a 'green' turn of mind, you might choose wave-power, contraception for men and women, electric cars, and so on.

In either case, the list is still vastly different from its ancient precursor. The only similarity is that the first Seven Wonders came largely from one culture – that of the Greeks – while the second set also comes from a single, though wider, culture – that of industrial western civilization. And for the same reason: both lists show the personal bias of their compilers!

Look at further differences from the past:

- None of today's Wonders will last for centuries. Existing examples will all be obsolete or worn out within a few years.

- None is a singleton. They are all classes of invention which have hundreds, thousands or even millions of examples.

- None existed a century ago. All are novel and continue to develop apace.

• Nearly all use energy or create it, sometimes in staggering quantities.

• Alarmingly, nearly all possess the capacity for massive misuse, accidental or deliberate.

One fascinating footnote – proof of how rapidly the world is changing – is that a similar inventory produced fifty years ago would have differed in yet another way. Most items on it would have been full of moving parts, unlike the original Wonders. Now, as the science of electronics advances, fewer and fewer bits that revolve or oscillate are involved.

Consider the two inventions which I argue were critically important in bringing us to where we are today. Clocks – perhaps for centuries the ultimate in moving parts – are now digital. Artificial power generation – which began with the beam-and-con-rods Newcomen engine – may soon advance to the stage of nuclear fusion. Here the heat source is contained in a static electromagnetic bottle, and heat itself could be converted directly to electricity – without the use of a rotating generator – through the medium of a magneto-hydrodynamic generator.

Inventors have to stand back and take a view of the way the world is developing. It is their task and privilege to find the pattern and move with it.

CHAPTER FIFTEEN

Ideas Have a Time to Be Born

'There is a tide in the affairs of men . . .' So there is, and it must be caught neither early nor late. The story of invention is filled with failures by people who either got ahead of the technology or lived in societies which were not prepared to take their ideas on board. The classic case is just that. It comes from ancient Greece.

Hero of Alexandria

There are a lot of misconceptions about Hero. His reputation as a builder of gadgets exists partly because of our idea that there was very little design engineering in the ancient world. This impression is false, as can be seen by looking at Hero's own written works. He flourished between AD 50 and 120 and was therefore familiar with Roman engineering as well as Greek. The subjects he wrote about included *Pneumatics*, *Automata*, *Mechanics*, *Mirrors*, *Measurements* and *Catapults*.

Mechanics is a reliable account of gear ratios, pulleys, distribution of loads, levers, screws, cranes and presses, and Hero's treatment of these matters is severely practical. His books on *Catapults*, which show military machines as unexpected as crossbows, are equally solid. He also produced such diversions as a penny-in-the-slot machine which dispensed a handful of holy water to religious worshippers, and a model temple with doors that opened and closed, apparently by magic.

When Hero's 'toys' are seen against the background of his knowledge, it becomes clear that he was using a brilliantly able mind to play around with ideas, in the same way that Leonardo did over 1000 years later. Most of his inventions are not for fun, but are strictly utilitarian, and include:

- a screw cutter for olive-oil presses

- a fire engine which actually pumped water
- an oil lamp with an automatically trimmed wick
- a hodometer, an instrument attached to the wheels of vehicles to measure how far they travel. The Romans used one version, which may have been Hero's own design, as a taximeter.

The rim of the wheel of the taxi (horse-drawn coach would be a better description) turned a rod which, through the action of a linked set of gears, slowly revolved a level disc, set with holes around its edge. Each hole contained a pebble. As the disc turned, the pebbles came one by one over the open mouth of a tube, down which they fell into a collecting tray. The numbers of pebbles discharged during the journey indicated the distance travelled.

Hero lacked a powerful prime mover to drive his machinery, but he got nearer to it with his steam turbine than anyone before Newcomen, 1600 years later. He described it as 'a ball which spins around on a pivot when a cauldron is boiled'. He let steam pressure build up in the cauldron, which was sealed, and ran the vapour through a pipe plumbed into the axis of the hollow ball. The steam then escaped through the nozzles of two bent and opposed tubes. This created a reaction thrust which made the sphere itself spin. It's the same principle as a jet engine.

Hero must have faced horrible problems in attempting to get any power out of this turbine, especially in generating steam at high pressure and sealing the ball against leaks at its axis, where the steam feed-pipe came in. He knew of all the devices needed to make a steam engine. He used a force pump in his fire engine, and valves in a syringe which he devised. His record for ingenuity and constructional skill suggests that he could have gone much further than he did.

Another device along the same lines was a machine which used wind power to play an organ. A wheel with projecting vanes (like a windmill), revolving in the breeze, turned a rod. With each revolution the rod knocked down a lever which, in turn, lifted a piston in a barrel. When the rod moved further, it slipped off the lever and the piston fell under its own weight. This pushed air into the organ and made it play. Once more, Hero did nothing to scale up the idea and use it to produce power in useful amounts.

I don't think Hero failed for technical reasons. More likely, there was no requirement for mechanical power in a society that could rely on the use of slave labour. Cheap muscle power was a much easier solution to any immediate problem.

Hero faced no demand for these inventions. They died with him. He was before his time.

Rule Two: Don't be before your time

Like Hero, many great innovators produce idea after idea. They refuse to be confined to a single development but carry on inventing, regardless of success or failure. If you accept my thesis that innovation comes from an attitude of mind, this is not as surprising as you might think.

Oliver Evans

Evans followed in the steps of Hero to build the first high-pressure steam engine in the USA, and later claimed to have had the idea for steam carriages in 1773, twenty-eight years before the Englishman Richard Trevithick. Evans was no rogue, or fool. He persuaded the assemblies of Pennsylvania and Maryland to give him an exclusive licence to profit from his inventions.

He made a high-pressure steam engine which worked at thirty rpm and installed it in his steam dredger, the Orukter Amphibole (wonderful name) on Pennsylvania's Schuylkill River. The dredger was an amphibian. It had power-driven rollers as well as a paddle-wheel, so it could move on land under its own steam as well as in the water. In 1804!

Although Evans went on to build fifty steam engines over the following fifteen years, the amphibian led nowhere. It was before its time. Another American, Robert Fulton, constructed the first successful steam boat in 1807 but fully amphibious vehicles had to wait until the present century. Evans was the first to design a factory which used no labour at all. It was the fully automated water-driven flour mill he and his two brothers erected in 1794 at Redclay Creek, Wilmington, Delaware. Once the grain had been loaded on to the inlet hoist, it remained untouched by human hand until it emerged as ground meal at the other end, bagged and ready for delivery. There was even a machine process which spread out the ground corn to cool before packing.

The inventor ran into trouble with his patent because, as his enemies pointed out, all the devices he employed – the endless belt, the screw conveyor, and chains of buckets – were known to Hero. This objection raises the interesting question: why didn't the Alexandrians build an automatic flour mill? What was new, of course, was Evans's overall concept. It was a brilliant anticipation of the twentieth century's attempts to make totally automated factories, which apart from cutting costs, save tedious and repetitious labour. But it was nearly two centuries too early.

The problem of innovating at the right time never changes and never goes away. Here are three stories from more recent times.

Shiro Okamura

The process of recording pictures and sound on tape has taken many years to develop. I remember the sensation created in the television industry by the first Ampex machines when they appeared in 1956. At long last, live programmes could be recorded and replayed instead of disappearing for ever as they were transmitted.

The invention was so outstanding for the world of professional broadcasting that I don't think it crossed our minds at the time that it would ever be possible for amateurs to record TV programmes *at home*. But it occurred to Shiro Okamura. By 1959, he had found a way to compress magnetic and video information so that a comparatively short length of recording tape could carry an enormous amount of information.

Unfortunately, domestic video recorders did not hit the market until 1975, when Sony's Betamax and JVC's VHS machines appeared. Okamura's patent had expired. If he had patented a decade later, he would have become a multimillionaire.

Alan Blumlein

Another note from the archives of the entertainment business. Alan Blumlein was one of the most brilliant and prolific twentieth-century inventors. He died at the tragically early age of thirty-nine while working on radar during the Second World War.

When he was with EMI, back in 1931, Blumlein patented an idea for recording stereo sound on gramophone records. It included the key notion that twin sound channels are imposed on the same groove in the recording disc and that the sound is replayed through two separated speakers. But until the arrival of light plastics in the 1950s, stereo recordings did not take off commercially. Even if he had lived, Blumlein was too far ahead of his time.

Felix Wankel

Although Wankel was one of the most fertile engine designers of modern times, he did not go to Technical High School or follow an engineering apprenticeship. His family's capital was wiped out by German mega-inflation in the early 1920s, so he started work in a university booksellers. Innovation began early for him. He devised a method of storing books in squares, which allowed them to be stacked ceiling-high without collapsing.

In 1924, while at the booksellers, Wankel started his work on rotary engines. Theoretically they are simpler than ordinary internal-combustion engines, lighter, smaller, and capable of burning cheap, lower-grade fuel. The main problem is sealing the engine: if the seal is too loose, power is lost through gas leakage; if too tight, through friction. So for twenty years Wankel earned his living by concentrating on the problems of sealing orthodox engines.

In the 1930s he worked for Daimler-Benz and then BMW. He was invited by the German Aviation Ministry to join the Central Research Establishment in Berlin but preferred to stay independent. The government gave him his own workshop because, by then, Wankel had patented his ideas to seal engines so that the gases did not leak. These were the key to his later income from licences, because the patents covered all the possible applications of his ideas for sealing engines and compressors, either conventional or rotary. His crucial idea was to use the gas inside the engine to apply pressure to its seals, instead of using springs. It was elegant and simple.

In 1954 Wankel showed the NSU company his drawings of a rotor with three convex sides, rotating in a double circle. In other words, the rotor was a bulbous triangle in a chamber shaped like a figure eight. Wankel said that he got the idea after eating too much Christmas pudding!

By the mid 1970s there were nearly 900 patents associated with the Wankel engine. NSU is said to have received more income from patent fees and royalties than any other German firm, nearly a tenth of the income for patents and know-how received from abroad by the whole of Germany.

NSU produced their first Wankel engine car in 1963 – a variation of the NSU Prinz. The ignition failed, and the sealing didn't work properly. The engine had never been properly tested. In the stop-start conditions of heavy traffic, the apex seals became brittle and broke up. Worse, no mechanic could handle repairs to a rotary engine, especially changing the seals.

This problem would not have mattered if the next car had been reliable. But the Ro80 also failed. Instead of soft-carbon rotor seals circling inside a hard outer surface, the seals were made of cast-iron alloy running on a silicon surface developed by Daimler-Benz. But the mileage was not much more than 10mpg. And the seals still broke down.

NSU eventually found the answer, using stronger alloys such as titanium nitride. It was too late – the commercial damage was done. Felix Wankel made a personal fortune but he got ahead of the materials technology of his day.

Now, with new materials, the latest version of his engine is doing well. The Commodore motor cycle has been highly successful.

There's a footnote to the Wankel story which shows he was not the first to be before his time with rotary engines. In 1908 an Englishman named Umpleby built an engine using a similar principle, with only one piston and three combustion chambers. It also failed.

Charles Babbage

One more story is worth telling at some length because it both exemplifies nearly all the problems an inventor can face, both then and now, and illustrates perfectly Rule Two: Don't be before your time.

Fellow academics didn't appreciate Charles Babbage's ideas; government was uncomprehending because practically no senior politicians had any scientific knowledge; costs roared hopelessly out of control because, if they had been known at the beginning of the project, it would never have got off the drawing-board.

Sounds familiar? Nothing has changed, has it? Babbage was consistently ahead of his time, one of that tiny group of polymaths who excel at everything they touch. He was charming, dynamic, led an intensely active social life and mixed easily with anyone in the land, from workmen on the shop-floor to members of the House of Lords. It is ironic that he is best remembered as the pioneer of computing, an enterprise in which he totally failed, although there has always been disagreement about what went wrong with his projects for Difference Engines and Analytical Engines.

Charles Babbage (1792–1871) trained as a mathematician at Trinity College, Cambridge, the university where he later became Lucasian Professor of Mathematics from 1828 to 1839. At the age of twenty-five he was elected a Fellow of the Royal Society. These were the only acts of official recognition he ever received in the land of his birth.

Babbage conceived the idea for his first Difference Engine ('Machine' would be a better word) when inspecting the navigation tables used by sailors in the 1820s. These were full of errors, partly because it was only too easy for the human calculators who worked out their huge strings of numbers to make arithmetical mistakes, partly because it was even easier for printers to copy the numbers incorrectly when preparing them for the press.

Babbage's friend, John Herschel the astronomer, compared the tables to hidden rocks at sea. There was no telling how many ships were sunk because of them. 'Why not?' asked Babbage. Why not devise a machine which would not only do all the repetitive calculations automatically, without error, but print the results as well? The tables would then be perfect; they could only be perfect. This task, and extensions of it, was to occupy him on and off for fifty years. Why did he fail?

Was the Invention on the Wrong Lines?

I won't outline the mathematical basis of the Difference Engine (a splendid display at the Science Museum in South Kensington explains it) but there has never been any doubt that Babbage's approach was right in principle. A modern version of one of his Engines works perfectly.

The first machine he designed contained 25,000 parts, and would have been eight feet high, seven feet long and three feet deep when completed. Part of the engine (with 2000 parts) was assembled in 1832 after eight years' work. It still exists in the Science Museum and has that weird beauty belonging to machines designed and made to high precision. It is a powerful scientific image today, and as a product of the first half of the nineteenth century it is quite extraordinary.

Did Lack of Money make the Invention Fail?

So it has been claimed. Babbage sought no profit from the Engine himself; he wanted the nation to have its benefits and he persuaded the government to back him. He fought strenuous and exhausting battles to get money from the State once it had been promised him. Leading politicians such as Sir Robert Peel – who was quite incapable of understanding what Babbage was trying to do – wanted only to get rid of this importunate inventor.

But the Duke of Wellington, when Prime Minister, backed Babbage and acted as his product champion. In the end, the government contributed over £17,000 towards the project. That's difficult to convert into modern terms but a sum approaching a million pounds would not be far wrong.

Was Babbage Swindled by his Manufacturer?

In 1991 the Science Museum unveiled the first-ever working version of Babbage's Difference Engine No. 2. This was designed in 1874 (long after No. 1 was abandoned by the government) and also never perfected or built. No. 2 gained from Babbage's early experience. Its design was simpler than No. 1, with only 4000 parts, and although it was eleven feet long and seven feet high, it was more elegant. It would have been easier to construct to Babbage's design tolerances because the state of engineering had improved hugely in the twenty years since No. 1.

The Science Museum faced many difficult problems in bringing the Engine to life after a 140-year gap, and it's a wonderful story in its own right, paralleling some of Babbage's own difficulties. Eventually, a month before the 200th anniversary of Babbage's birth, the Engine clunked into successful action at a project cost of just under £300,000.

The man contracted to make the parts for the original Engine was Joseph Clement, one of the leading figures of his day and a member of the great group trained by Henry Maudslay (see page 179). The quality of work Clement supplied was superb, with parts repeatable to an accuracy of two-thousandths of an inch, an outstanding achievement for the 1820s.

Was Babbage overcharged by Clement? On the one hand, Clement was undertaking entirely new work and delivering what Babbage required. On the other, he had produced only half the number of parts required by the time the project was closed down. Again, that half represented 12,000 parts compared with the 4000 used in the modern Engine. Clement's (adjusted) costs match this difference: they were almost exactly three times higher than the costs for No. 2. Calculation here can only be rough, but it looks as though Clement was behaving moderately honestly.

Was Babbage Bad at Dealing with People?

Babbage certainly had little luck with most politicians, hitting a period when classically educated know-nothings like Lord Melbourne were lackadaisically in control of government. But he had no such trouble with the intensely practical Duke of Wellington, who cared not a fig for Greek or a gentleman's education. The Duke instantly saw the potential military value of improved navigation tables.

The accusation against Babbage has more to do with his arguments with Joseph Clement. He worried about the bills Clement was submitting and was only too conscious that Clement had the whip hand because, under the law of the time, he owned the tools which Babbage himself designed to make the engine.

The relationship did not run smoothly and matters came to a head when Babbage wanted Clement to move his workshop from the Elephant and Castle nearer to Babbage's house in Dorset Street, four miles away. After much chaffering Clement refused to move, despite an offer of compensation and even though most of his business was working for Babbage. For his part, Babbage refused to pay any more of his own money, telling Clement to bill the Treasury direct. Clement fired his workmen and work on the project stopped.

Babbage – whose friends regarded him as the most amiable of men – does not seem to have been at fault personally, and Clement was a prickly customer. But Difference Engine No. 1 was dead and that was a loss that could not be recovered.

Was Babbage too Dilettante?

Babbage certainly dabbled in many projects which did not come to much. He proposed a submarine with an open bottom which would let a diver attach mines to underwater targets; he played around with ciphers, which he was brilliant at solving; he designed a hydrofoil that could not travel because there was no motor powerful enough; he worked on a games-playing automaton but never built it; he designed an ophthalmoscope which he did not sell.

But the inventor had genuine achievements to his credit, especially occulting lights for lighthouses and communications (taken up, not in the UK, but in the USA and Russia, which used them against Britain during the Crimean War). He worked out that the cost of collecting and stamping letters for different sums depending on how far they were to travel was more expensive than a simple, flat-rate charge. This led directly to the introduction of the penny post in 1840.

Babbage produced actuarial tables of life expectancy for insurance companies. He designed the first railway track recorder. His on-the-spot studies saved Isambard Kingdom Brunel's plans for the new Great Western Railway and were a forerunner of monitoring equipment today.

With the Difference Engines themselves, and their more advanced counterparts, the Analytical Engines, which were likewise never built, Babbage never gave up, though as he grew old his activities on their behalf weakened.

Was Babbage Ahead of his Time?

None of the suggestions made so far to account for Babbage's failure stand up very well, although there must be some truth in the charges that he mishandled the Clement problem and failed to realize how much Difference Engine No. 1 would cost to develop.

Was the true reason for his lack of success that he was ahead of his time? The accusation was levelled against him for many years, but now there is a tendency to acquit him of it. Is acquittal justified?

We need to clear our minds of the benefit of hindsight. Hindsight suggests that because Clement was able to supply what Babbage demanded, Babbage was not asking too much. Hingsight also proposes that because Engine No. 2 works today in the Science Museum, it would have worked in the 1840s. But Clement took eight years to produce just half the parts required for Engine No. 1. How much longer would it have taken to deliver the rest? Another eight? Is this a reasonable development period for any product that has to be financed with speculative money – and large sums of it at that?

Would Engine No. 1 have functioned properly, even if it had been delivered? Part of it worked well, but many smaller-scale devices do this, only to reveal new problems when they are fully assembled. Work on No. 1 did not include the printer, which was a fundamental part of the idea. That would have involved yet another manufacturing and testing period.

As for Engine No. 2 – yes, it works now. But when the experts at the Science Museum scrutinized Babbage's designs, they found some mechanisms missing and one that was unnecessary. One part of the machinery was unworkable. These oversights had to be corrected before No. 2 could be built. There's no doubt Babbage would have solved these problems if he had ever got round to constructing the machine but, again, they would have taken time and money.

The project suffered from the classic problems of a pioneering enterprise. It was a one-off (always expensive) and it pushed at the frontiers of technology.

We owe Babbage a debt for the improvements in mechanical engineering which his project created. By the 1840s, engineering companies like Joseph Whitworth's were turning out products made to tolerances of one ten-thousandth of an inch.

But, sad to say, there is no direct line between Babbage's work and modern computers. He *was* ahead of his time and that's what sank him.

CHAPTER SIXTEEN

Self-help

Many famous innovators were self-educated, or trained in areas unconnected with those where they achieved fame. Experience demonstrates that despite lack of formal qualifications, if inventors are motivated and keep learning, they find success.

I've already mentioned Felix Wankel, who started work in a bookshop but went to night school to acquire a technical foundation for his developing interest in engines. There are many more like him.

Rule Three: Experience is never wasted

• Joe Brammer (1749–1814) was a Yorkshire plough boy. He injured his ankle when he was sixteen and, while lying in bed and filling the empty days with his hobby of wood carving, he decided to become a cabinetmaker. After he had served a seven-year apprenticeship there was no work to be had, so he walked all the way to London to look for fame and fortune.

There he found employment and, in his spare time following a second fall, designed the first really effective flushing lavatory. Not only that, but from his carpenter's wages he saved £120, enough money to take out a patent. The lavatory gained fame as 1177, from its patent number.

Joe was to become one of the greatest of engineers, but not under the name of Brammer. He faced an absurd class system which sneered at him because he was both a provincial and a working man, so he changed the spelling of his name to Bramah. The long 'a' sounds of Bramah were posher than Yorkshire Brammer. But working people never forgot where he came from. His developments in the fields of hydraulic and mechanical engineering were so outstanding that his name passed into their language. My co-author tells me that when he was serving an apprenticeship in the Midlands, more than 200 years after Joseph Bramah's birth, a particularly well-made piece of engineering was still known as 'a real Brammer'.

What's noticeable about Bramah's early years is not only his lack of formal training, but his determination. He found a new career when a crocked leg took him away from the plough; he decided to walk to London when there was no employment for him in Barnsley; from a very small wage, he saved to build the capital which enabled him to pay for his patent.

• The American Edwin Land deliberately decided not to finish his formal education. He was in a favoured position. His father was a landowner who also ran a scrap-iron business; the family had money. Land was a brilliant student and he entered Harvard to study physics at the age of seventeen. But he never completed his degree studies.

Land was still a freshman when, in 1926, he visited New York. Wandering down Broadway, he became intrigued by the glare of lights from competing theatre and club advertisements and wondered if there was any way of eliminating it. He began investigating an idea to polarize light by orienting small crystals of a chemical called iodoquinone sulphate in a sheet of transparent plastic. It was so fascinating that he dropped out of his degree course; it wasted too much of his time!

Land eventually produced Polaroid, which was ideal for sunglasses and of scientific importance in colloid chemistry. Success in business did not stop his urge to innovate. In 1941, while he was on holiday in New Mexico with his family, he took a photograph of his three-year-old daughter. She asked him why she couldn't see the picture straight away? Land went for a walk around Santa Fe. 'Within an hour the camera, the film and the physical chemistry became clear,' he later recalled. Seven years later, after many experiments, his Polaroid system of instant photography went on sale.

Land may have given up formal education because he had more urgent things to do, but he never stopped learning. He became a leading researcher into colour vision, publishing papers and giving lectures on the theory of colour constancy (which attracted considerable hostility from established scientists). He may have been wrong, but in 1957 Harvard awarded its drop-out an honorary doctorate, and he died a multimillionaire.

• Clive Sinclair refused to go from school to university, although he was a whiz at physics and mathematics. 'I already knew a lot about electronics,' he was quoted as saying. 'I also knew that if I went to university I couldn't just study electronics, which interested me; I'd have to do electrical engineering.' This sounds very like Edwin Land's decision to drop physics and, equally, it did not stop Sinclair innovating. His pocket calculator, which appeared in 1972, was one third the size of any competitor and cost half as much. He achieved this by finding a way to reduce the power input demanded by the calculator chip. The Sinclair Z-80, which started more Britons on home computing than any other machine, followed in 1980.

Sir Clive is another innovator who never gave up learning. Years after he first turned down the chance to go to university, he went to Cambridge and took a degree in economics.

All three innovators, spread over 200 years of history, missed some or all of the stages of formal education: Bramah because he had no choice, Land and Sinclair because they judged it not necessary. Their success does not mean that innovators can afford to do without practical or theoretical knowledge of the areas in which they are working, but it does show that the desire to invent can arrive before academic qualifications.

The threesome have other characteristics in common: a hard, driving will to succeed and an ability to keep on producing fresh innovations. Without these qualities – which are those needed to take a bright idea from the 'light bulb over the head' stage through to commercial success – no innovator can hope to profit in business.

We know the wry saying 'It's not what you know – it's who you know', with its implications of favouritism and corruption. Forget the bad side – no invention can succeed in the market-place simply because the innovator is a prominent member of some society of businessmen, or has a sister who is a buyer for a chain store.

At the purely technical level, it's important to know what the competition is doing; on the business side, to find markets or discover what consumers are demanding. Only personal contacts can give this information. It's no good living in an ivory tower, isolated from the human race, hoping that the world will beat a path to the bottom of your spiral staircase. That's not the way the world wags. The innovator has to get out and about, to keep learning.

Rule Four: Join a network

I cannot prove that knowing other people leads to those small, unquantifiable, advances in innovation that produce success. But we all know that life depends on subtle points of personal interaction, messages passed along chains of communication that may involve four or five links, and on being in the right place at the right time. On a broader scale, there are networks of innovators whose relationships changed history. If it worked for them, why shouldn't it work for anyone?

Let's return to Joseph Bramah. Without intending it, he became the founder of a world-changing group of engineers. For over a century, anyone who wanted to stake a claim to fame in engineering could plug into the network by training in the workshops and factories of its members. After the success of his flushing lavatory, Bramah took on a young man called Henry Maudslay as chief assistant, and set up a factory where he developed the hydraulic press.

The press exerted forces up to several thousand tons. It was strong, but delicate in its touch. It could tap bales of cotton into

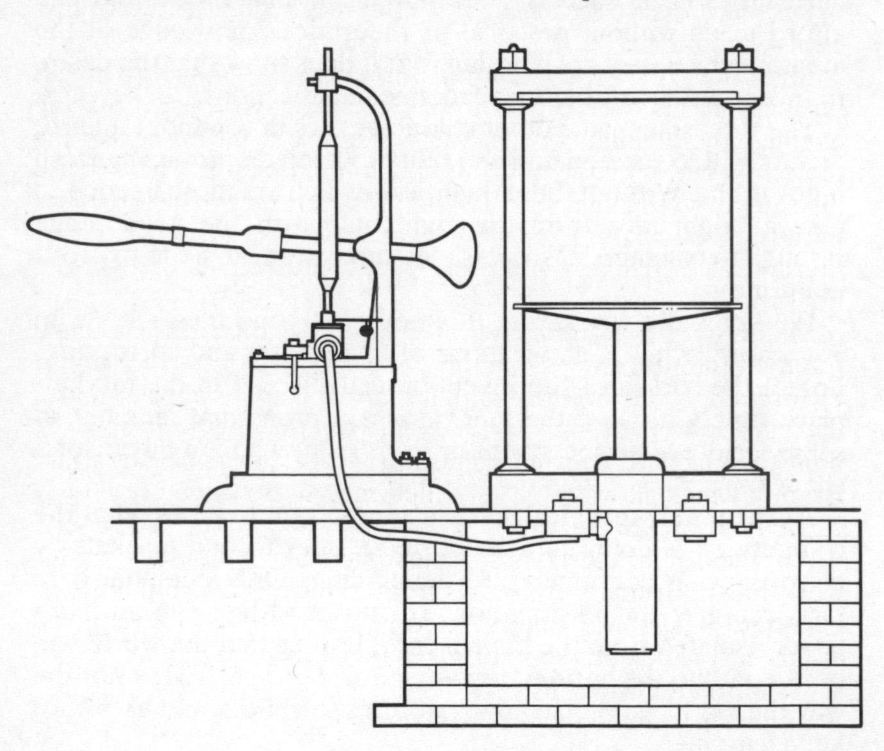

shape or forge metal ingots – where its steady pressure gave a precision lacked by steam hammers. It made the girders for Robert Stephenson's railway bridges and its development as a hydraulic jack launched Brunel's gigantic ship, the *Great Eastern*.

Maudslay made superior tumbler locks to Bramah's design, and in the 1790s Bramah offered a prize of 200 guineas to anyone able to pick one of them. It was not until 1851 that an American locksmith, Alfred Hobbs, brought off the feat – and even then, he had to cheat to do it.

Bramah invented a machine which printed banknotes with sequential numbers; a beer engine; a fountain pen; a planing machine; a hydro-pneumatic suspension for coaches rather like Alex Moulton's system for cars in the 1950s. He was an incomparable innovator.

Part of Bramah's reputation rested on Henry Maudslay. His hydraulic press would not have succeeded without Maudslay's invention of the leather cup washer. This runs round the circumference of the piston, and under hydraulic pressure it bulges against the walls of the cylinder, forming a perfect seal; when the pressure is released, it collapses so that the piston is easily withdrawn.

Maudslay (1771–1831) outdid his master after he set up in business on his own. Apart from pioneering work with marine engines, the highlights of his innovating career were perfecting the screw-cutting lathe and the bench micrometer. This second instrument could measure to one ten-thousandth of an inch. Maudslay called it 'The Lord Chancellor' because, whenever there was a dispute about measurement, the micrometer was the final court of appeal.

Maudslay introduced the idea that castings should have rounded internal angles. 'Always avoid sharp angles in castings or forgings,' he would say, holding up his hand and pointing to the curved web of skin between his outstretched fingers. He built the machinery for making ships' blocks, designed by Marc Brunel, which allowed ten unskilled men to do work previously carried out by 110 craftsmen.

Another of Maudslay's legacies was the 'graduates' of his workshops. Brunel's son, the great railway and ship builder Isambard Kingdom Brunel, stated in later life that he acquired all his early understanding of mechanical engineering at Maudslay's Margaret Street works.

Engineers trained by Maudslay included:

- James Nasmyth (1808–1890), inventor of the steam hammer,

which greatly increased the size of forgings without losing any precision. Nasmyth ran a foundry and locomotive works at Patricroft, near the junction of the Liverpool to Manchester railway and the Bridgewater canal. He made so much money that he was able to retire before he was fifty. But he didn't give up learning. He took up astronomy and surveyed the surface of the moon. At the Great Exhibition in 1851, his steam hammer was on display and he also won a prize for lunar map making.

• Richard Roberts (1789–1864) set up his own firm in Manchester after working for a time with Maudslay. He was a versatile inventor of planing machines, improved lathes, an electromagnet and a number of clocks. His most important innovation was to make Crompton's spinning mule self-acting. Local cotton manufacturers asked him to do it in order to counter a strike and he succeeded in four months.

Like Nasmyth, Roberts entered the world of steam-locomotive building and, in 1848, was again asked to help beat a strike, this time by skilled men working on a Welsh suspension bridge. He came up with a machine for punching holes at exact intervals in steel plate.

Unlike the remainder of Maudslay's 'graduates', he died in poverty.

• Joseph Clement (1779–1844) we have already met in connection with Charles Babbage. He was chief draughtsman for Bramah and then for Maudslay. These men's lives and careers interlock at many points. Clement followed up Maudslay's work on accurately cutting screw threads, and invented the tap with a square shank which can be knocked through a threaded hole instead of having to be screwed out – a device for which eight generations of workmen owe him thanks.

Clement remained in the London area, with works just south of the River Thames at the Elephant and Castle. Like Roberts and Bramah, he also made a planing machine.

• Sir Joseph Whitworth (1803–1887) worked for both Maudslay and Clement before moving to Manchester. He was in many ways the closest of Maudslay's disciples to his spirit. Most engineers built machine tools for themselves but Whitworth sold them to other manufacturers. He raised the standard of machine work everywhere. By the 1840s, his tools routinely operated to one ten-thousandth of an inch – a standard which could be measured by Maudslay's Lord Chancellor only thirty years before but not achieved.

Workshops had their own ways of turning out screw threads, which stopped them being interchangeable with others. Whitworth founded a uniform system of threads with a constant angle of fifty-five degrees between their sides, and fixed numbers to the inch. It was standard throughout the former British Empire until 1948.

Whitworth was another believer in learning. He set up thirty scholarships to encourage mechanical engineers, and gave huge sums to colleges in the Manchester area.

I could continue multiplying membership of this illustrious engineering network, but I think I've made the point. Bramah trained Maudslay and Clement, Maudslay trained Brunel, Nasmyth, Roberts and Whitworth. Brunel employed a generation which became famous as railway engineers; Whitworth's factories and scholarships were responsible for many more.

Skilled men transferred easily from one part of the network to another. Nasmyth, Roberts and Whitworth – Maudslay's old boys – all had works and factories near Manchester. And the idea of learning, of bettering oneself, was not lost even in the Gradgrindian depths of industrial England.

Similar networks existed throughout the industrialized world. Sometimes it was a connection between ideas, of learning what was happening without the engineers themselves being in contact; sometimes they worked for the same company.

- The French innovator Jean Lenoir was first to succeed in obtaining power from combustion *inside* the cylinder of an engine, whereas steam engines used an *external* source of energy from burning coal or coke. He patented his gas engine and fitted it into a horseless carriage. The motor vehicle was born.

- Nikolaus Otto, a young German clerk, came across a newspaper article describing Lenoir's engine and decided to do better. He learned how to compress the mixture of air and fuel before it ignited near top dead centre, and he sorted out the use of separate strokes for expansion, exhaust and induction. The result was a small four-stroke engine. After a number of financial problems, he patented the improved engine for which he is famous in 1876. It was more efficient than Lenoir's engine, and so quiet by comparison that it was known as 'the silent Otto'.

- Meanwhile Gottlieb Daimler, who – rare among these innovators – trained as a mechanical engineer, had been struggling to devise a small, low-cost engine but with lack of success, although he too became interested in Lenoir's engine during a stay in Paris. He spent ten years in heavy engineering, where he met Wilhelm Maybach, and the pair eventually joined Otto's company, Gasmotorenfabrik Deutz, near Cologne. Otto was technical director but he left the process of perfecting his four-stroke engine very much to Daimler and Maybach.

• Maybach and Daimler set up their own company in 1882. They raised engine speeds from 300 to 900 rpm, and lightened the engine so much that it could be used in a motor cycle. Maybach, who was the son of a carpenter, made major improvements to carburation, gearing, fuel injection and timing, and in 1894 a Daimler car incorporating his ideas won the first international motor race.

The network was in operation again. First, with learning. If Otto had not heard of Lenoir, he might never have started work on his own engine. If Daimler had not seen Lenoir's engine in Paris, he might have given up his own ideas in that direction. Then, if Otto had not founded his own company, he might never have met Daimler. If Daimler had not met Maybach outside the internal-combustion-engine industry, Maybach would not have become an auto engineer.

The stories of the Bramah–Maudslay and the Otto–Daimler networks show how needful it is for innovators to help themselves, to learn, and to be in touch with the professional world in which they operate. They also need empathy, the ability to strike a rapport with others. If innovators continually annoy or upset professional colleagues and competitors, life can turn very sour indeed.

One further story underlines the importance of Rule Four: Join a network.

• Benjamin Thompson, Count Rumford (1753–1814) belonged to every network it was possible to join, and was thrown out of nearly all of them. Previous generations called him a soldier of fortune; and yet, most confusingly, he was also a considerable scientific figure and innovator.

Marriage to wealthy widow – One Thompson grew up in Britain's American colonies at the time when they were flexing their muscles in readiness for the War of Independence. At the age of nineteen, starting early on his career of enlightened self-help, he married Sarah Rolfe, a wealthy widow fourteen years older than himself.

One of the UK's most successful and professional private innovators – Ron Hickman with his Workmate. (*Picture: Robin Bootle*)

The rewards of invention – Ron Hickman at home in the Channel Islands with his 1931 Cadillac, which once belonged to a maharajah. (*Picture: Robin Bootle*)

The rewards of invention – on a smaller scale. The bathroom which Bob Symes was given by Ideal Standard as an ex-gratia award for his invention of the smell-free lavatory (page 128). The lavatory on view is fitted with the invention, of course. (*Picture: Robin Bootle*)

Another Bob Symes invention. On the left, a generator; in the centre, a set of car batteries; above the batteries, a box holding the inverter that changes direct current from the batteries into alternating current which runs low-powered equipment in the home. Further circuitry tells the generator to start up when demand for electricity rises above a certain level. (*Picture: Robin Bootle*)

The centrally pivoted cat flap in use (page 135). The cat comes out on the right hand side and enters the house again through the left. (*Picture: Robin Bootle*)

The Monodraught chimney. Because it does not suffer from downdraught problems, it can be set well below the highest point of the building. Note how much higher the conventional chimneys stand. (*Picture: Monodraught Flues Ltd*)

The evolution of the Monodraught chimney into forms which are protected by Registered Design. The design – space age or traditional – can now be used as the architectural high spot of the building. (*Picture: Monodraught Flues Ltd*)

The Great Pyramid – only survivor of the Seven Wonders of the ancient World. No moving parts, no use of energy, for ever unchanging and totally different from today's Wonders of invention. (*Picture: Jeremy A. Horner/ The Hutchison Library*)

Reg Crick, who rebuilt Charles Babbage's Number Two Difference Engine for the Science Museum in London, shows its scale against a 1990s laptop computer. (*Picture: Daily Telegraph/Ian Jones*)

Inset: Charles Babbage. Daguerrotype portrait by Antoine Claudet, taken around 1840. (*Picture: The National Portrait Gallery*)

The inventor as master of all he surveys. Richard Trevithick, inventor of the railway locomotive, painted by John Linnell in the 1820s. (*Picture: The Science Museum/ Science and Society Picture Library*)

'Catch me who can.' Trevithick's last locomotive, built in 1808, running on a custom-built track near what is now Euston Station. Drawing by Rowlandson. (*Picture: The Science Museum/Science and Society Picture Library*)

Nikola Tesla in his laboratory at Colorado Springs 1900. This brilliant piece of self-publicity is a double-exposure. Tesla took up his position only after the ten million volt electrical discharge ended. (*Picture: Nikola Tesla Museum, Belgrade*)

Another example of Tesla's genius as a publicist. This picture is not faked. There really are no wires connected to his invention, the world's first fluorescent lamp, which is remotely powered by a high-frequency electric field. (*Picture: Nikola Tesla Museum, Belgrade*)

Inventors of the future (1). Lucy Porter from Bath High School with her swing exerciser for disabled children. It won her the titles of Young Engineer for Britain and Rotary Young Inventor of the Year. (*Picture: Rotary International in Great Britain and Ireland*)

Inventors of the future (2). The Project Volta team from St Richard's School, Bexhill, whose vehicle broke the world land speed record for light electric cars. (*Picture: Tony Hillyard for Young Engineers*)

Hasty departure from America The marriage brought him into social circles surrounding the British governor of New Hampshire, who gave him a commission as a major in the militia. Thompson tracked down army deserters and spied on his rebellious fellow Americans. The country became too hot to hold him and he deserted his wife and child, to flee to London.

Within four years, Thompson's social connections made him an Under-Secretary of State. In his plentiful spare time, he researched the explosive power of gunpowder and a system of signalling at sea; he was elected a member of the Royal Society.

Hasty departure from England There were rumours that he was involved with a French spy called LaMotte, who was caught holding secret naval documents. It was politic to return to America, where he became a colonel in the British army.

Hasty departure from America When the Revolutionary War ended in American victory, Thompson could not stay in the disaffected former colony. And, although he had been knighted by that famous judge of character, King George III, it wasn't possible for him to live in England either: too many people were gunning for him.

Scientific work in Germany Thompson set off for Bavaria, where he ingratiated himself with the Elector and stayed for eleven years as Minister of War, Minister of Police and Grand Chamberlain. His services were so admired that in 1791, a year after the death of his wife, the unfortunate and abandoned Sarah Rolfe, he was ennobled as Count Rumford of the Holy Roman Empire. (The Empire itself was abolished only fifteen years later but, for once, this does not seem to have been Thompson's fault.)

In Bavaria, he grew interested in the problem of heat and attempted to work out the connection between heat and work. Conventional thinking at the time suggested that heat was some kind of liquid. Thompson studied the way it was produced when gun barrels were bored. He rightly concluded that the movement of the borer was converted to heat and that, therefore, heat was a form of motion, or energy as we would say today. He even tried to work out how much heat was produced by a known amount of work.

Diplomatic but compulsory departure from Bavaria In Germany, Thompson made many enemies, who got rid of him by sending him back to England to represent Bavaria at George

III's court. There he was refused recognition because he was an Englishman and it was treasonable for him to act for a foreign ruler against his own King. Rumford had a genius for inventing problems as well as solutions.

Scientific work in England Once again, spare time became his, and he invented the Rumford stove. An efficient way of burning fuel in a cooker which also heated the building, it was a precursor of the Rayburn stove. His compatriot Benjamin Franklin had devised a similar idea a few years earlier.

Thompson enlisted the help of the wealthy amateur scientist and botanist Sir Joseph Banks to found the Royal Institution (RI) in 1800, with Humphry Davy as its scientific lecturer. The idea was more than successful. Since Thompson's day, heads of the RI have included Michael Faraday, the Braggs – father and son – and Sir George (now Lord) Porter, Nobel prize winner for chemistry.

Aggravated departure from England The directors of the RI grew tired of Thompson's dictatorial manner. He went off to France, which, under Napoleon, was at war with England, although that didn't worry Thompson one bit. In Paris, he studied heat generated in combustion and invented an improved lamp. French scientists were generally cold in their attitude to him – perhaps they saw him coming.

Marriage to wealthy widow – Two The great French scientist Antoine Laurent Lavoisier was guillotined during the French Revolution – as a tribute to equality and fraternity? – Thompson married his wealthy widow. The couple had their first row the day after the wedding and separated within four years. Thompson later claimed that Mme Lavoisier was such a crow that her former husband could count himself lucky to have had his head chopped off.

Respectability after death Thompson had every advantage in terms of scientific intelligence and connections – using the network. He was also egotistical and overconfident, irritating nearly everyone he met to the point where they couldn't stand his presence any longer.

Posterity has been kind to him. He won approval for refusing to patent his inventions because he wanted them to be in the public domain. In the USA, he became respectable because he left most of his estate to found a professorship in applied science at Harvard. In the UK, he is remembered by the Rumford

Medal, one of the highest awards the Royal Society can bestow.

I have a different view. Thompson led a life which betrayed the value of networking, and he paid the penalty by being hounded from one country to another for thirty years. He was a good scientist who could have achieved much more had he not made himself a semi-permanent fugitive.

Rule Five: Prove your invention works

There is a tendency among writers about invention and the history of technology to credit an invention to the person who first thought of the idea. This piece of academic nonsense needs stamping on. Even someone as close to success as Charles Babbage with his Difference Engine does not merit the title Grandfather of the Modern Computer (as one respected work of reference lying in front of me describes him) because no consequences flowed directly from his work. Hero of Alexandria is not the Father of the Steam Turbine for exactly the same reason.

Proof of invention lies in taking the product to the marketplace; then it can truly be attributed to the innovator. There is a second class of inventions which are not successfully sold, but where a working model exists to prove that the idea was more than a nice thought in the innovator's mind. Outside these two categories, no claim to novelty can be admitted.

The working model is crucial. This is why the claims often made about Leonardo da Vinci's deeds as an innovator are greatly inflated. Leonardo left thousands of drawings of machines, now grouped in various collections in Italy, France, England and Spain, which are thought to be only a quarter of his output, the rest being lost. They cover battle machines, hydraulic machines, flying machines and other mechanical devices, including a disproof of the existence of perpetual motion (my heart warms to the man).

I can see why exaggerated claims are made for Leonardo. Who can resist the small drawing which seems to depict a Renaissance bicycle? The machine-guns and tanks, the ornithopters and primitive helicopter screw? Not to mention the artistic brilliance of the sketches.

Museums in Italy and France contain models, laboriously made on the basis of Leonardo's drawings, that make us wonder at the fertility of his mind. But these models also give away the truth. They had to be built from drawings because, as far as we know, the inventions themselves never saw the light of day in

Leonardo's time. And – note this – the models don't work; the machine-guns don't fire, the planes don't fly. They were solid inventions only in Leonardo's mind.

Moreover, Leonardo did not publish in his own lifetime. Nor were his ideas broadcast after his death. He left his notebooks to his faithful companion Francesco Melzi. Melzi saw it as his duty to guard the maestro's work and refused to let anyone have access to the notes.

It was only in the twentieth century that scholars started to interpret Leonardo's ideas in practical terms. The bandwagon began to roll with a Fascist exhibition at Milan in 1939 which glorified Leonardo as typical of Italian genius. Since then, the movement to describe him as a great innovator has become almost unstoppable.

Let's be clear. Leonardo did not make working models. He did not publish. He even described his ideas in mirror writing, inverted right to left, which made them more inaccessible. He has no claim to priority with tanks or aeroplanes. No working model – no invention.

CHAPTER SEVENTEEN

Work with the System

There are areas of life where a distribution network or operating system must be constructed before an invention is usable. Transport is the obvious example: there may be several hundred models of car on offer to the consumer, but they all have to travel on the same roads. Communications is another: equipment attached to the end of a telephone line may be as varied and ingenious as answerphones, modems or fax machines (not to mention the telephone itself) but it employs the same transmission lines and exchanges.

In the area of personal computing, there are many competing programs available for word processing, graphics, spreadsheets, databases and so on. They nearly all run on machines using IBM or Apple Macintosh operating systems.

What decides the choice of network or operating system itself? Being first in the field is most of the answer. It's rarely possible for newcomers to displace the people in possession (although this chapter contains the story of Nikola Tesla, who did just that). History shows that heroic though challengers may be, they rarely win.

Rule Six: Don't become a dead hero

Perhaps an existing system is inefficient or has simply grown up higgledy-piggledy; a better method may be easy to devise. That doesn't mean it is sensible to fight giant companies or established practices.

When Isambard Kingdom Brunel won the contract to design the Great Western Railway (GWR) – he was only twenty-eight! – he resolved that it should be 'the finest work in England', and the fastest. He created a line from London to Bristol where 70 out of 120 miles were on gradients of less than 1:1000. This was to help locomotives draw the trains with maximum speed. But speed had its own drawback: safety.

Brunel was a radical thinker who never let precedent get in his way. He decided the way to ensure safety was to set the rails farther apart than the standard gauge of four feet eight and a half inches. The existing gauge was completely illogical anyway. George Stephenson had chosen it simply because the trams in Killingworth Colliery, where he had been chief mechanic, used it. Brunel persuaded the GWR to adopt a gauge of seven feet. The change let trains run safely at high speeds and locomotives themselves became more powerful because there was extra room between the wheels to boiler them up.

Among Brunel's many other innovations were a new type of rail, and a bridge over the Thames at Maidenhead with the longest brick arches ever constructed (the subject of Turner's famous painting *Rain, Steam and Speed*, it stands to this day).

Brunel was an engineer of genius and his arguments in favour of broad gauge were sound. So why do we still travel on standard gauge, and why was the Great Western, after half a century of doughty struggle, forced to remove the last of its broad-gauge track in 1892? A Royal Commission appointed in the 1840s to decide the battle of the gauges agreed to a contest between one of Brunel's broad-gauge machines and two standard-gauge locomotives. Between Paddington and Didcot the broad-gauge loco averaged fifty miles an hour, pulling a sixty-ton load. Standard gauge was totally defeated.

Nonetheless, the Commissioners recommended that standard gauge should be adopted for Britain – a choice that affected future railways all over the world, which to this day run on the gauge of Killingworth Colliery's tramway!

The reason for the Commission's apparently illogical decision was that standard gauge *was* the established system. Over 2000 miles of standard track were already laid, while the GWR had less than 300. It was also easier to reduce gauge than enlarge it. The GWR lost its case. It wasn't possible for the gauges to live side by side because trains could not use two systems if they were to carry passengers and freight to all parts of the country. For a time, people and goods were trans-shipped at railway stations where the different systems met. Then a third rail was added to GWR track to allow standard-gauge trains to run on it. It was the beginning of the end.

If Brunel had been first with the broad gauge, he would have succeeded. As it was, he and his system became dead heroes – right, but definitely dead.

The steam car has long been touted as a sensible alternative to vehicles with an internal-combustion engine (which I'll call petrol cars, for the sake of brevity). In recent years, I have been able to look at a pilot steam vehicle commissioned by the Australian government. It had a three-cylinder engine, a slow reverse gear, could stand at rest in a traffic jam for twenty minutes before running out of steam, and went like a sports car.

It used totally modern technology and weighed less than half a ton. Its electronically controlled steam generator ran on paraffin or kerosene.

Steam cars have a long and mixed history. They twice held the world Land Speed Record but singularly failed to equal the developments which turned the unreliable petrol car into a vehicle which could be driven safely by the non-mechanically minded. None have been sold as production models since the 1930s.

Yet they have many virtues. Imagine that steam cars were on the road today. When it comes to clean exhaust emissions, they have a head start over petrol cars, for carbon monoxide and unburnt hydrocarbons are a vanishingly small part of their discharge. They relieve pressure on non-replaceable oil reserves because they burn low-grade fuels. With up-to-date construction methods, they are almost silent. Think how the lives of city dwellers and people near motorways would improve if the roar of petrol engines vanished.

On top of these environmental gains, steam cars offer simplicity of control. Steam engines have a high starting torque, which means that they need neither clutch nor gearbox. Acceleration is smoother than with an automatic gearbox. But they do have disadvantages. One is their start-up time. Even a wait of thirty seconds while a flash boiler generates steam is too long for some drivers. Another is running costs. It's argued that because steam cars use low-grade fuel, they are cheaper than petrol vehicles to run, but this does not allow for the steam engine's lower thermal efficiency. Cynics suggest that if steam cars could be made to burn lawnmower cuttings, it would be only a matter of months before the Chancellor of the Exchequer introduced a chlorophyll tax!

But there's a more crucial reason why steam cars will never make a comeback, unless the petrol engine is banned first. The only people who can make them in quantity are the existing car manufacturers. But that would involve them in heavy initial investment, and the new cars would challenge their existing products. It's not going to happen.

In the 1960s there was an attempt by an American, William Lear, to produce a modern steam car. He spent several million dollars on the project before giving up. Another dead hero. I fear that the Australian steam car I mentioned at the beginning of this section will go the same way.

Rule Seven: Operate the system

Avoiding being run over and squashed flat by big business is one thing; persuading those businesses to take up an innovation and push it forward as if it belonged to them is another. The inventor needs to understand the patent system. Only then is it possible to find a backer with industrial clout, a product champion who can make the difference between success and failure.

This was how Nikolaus Otto became commercially successful. As I've already described, Otto got the idea for his internal-combustion engine from the work of Jean Lenoir. Although Lenoir had priority with his atmospheric gas engine and sold a number of them, it was Otto's improvements and changes to the idea which created the first working internal-combustion engine. This wasn't simply the luck of the draw. Otto gained a patent for his first engine, which was atmospheric, like Lenoir's. This attracted the attention of industrialist Eugen Langen, who helped him set up a firm called Nikolaus August Otto and Company.

Otto and Langen exhibited the atmospheric engine at the Paris Exposition of 1867, where it won a gold medal because it was nearly three times more efficient than machines shown by Lenoir and other competitors. But production could not keep up with sales, and Otto would have been in severe trouble without Langen's support. They raised more capital to set up a new company, Gasmotorenfabrik Deutz, which built a factory near Cologne in 1869.

By 1877, Otto had produced a fully operational four-stroke engine running on gas. It was a tremendous success. But his patent was contested and finally, in 1886, invalidated because a Frenchman called De Rochas had patented a similar idea in 1862.

De Rochas described the theory, with no attempt to solve the practical problems that Otto had overcome, but that was enough to destroy Otto's claim. Did Otto know of De Rochas' patent? We'll never know. It didn't matter. With Langen's money to tide him over rough patches in the early days, and the following sheer commercial success of his engine, 30,000 of which were sold in the first ten years, Otto moved smoothly forward to produce a design which ran on petrol instead of gas.

The modern car engine was born. Otto won the market for internal-combustion engines and Lenoir was driven from the field. Otto's patent (however faulty) and his product champion won the day.

Another innovator in the engine business was Rudolf Diesel. He, too, profited from observing the successes of others. Otto gained inspiration from Lenoir's engine; Diesel took his from Otto's.

In 1892 he patented a high-pressure engine and wrote *The Theory and Design of a Rational Heat Engine*. An important company – Maschinenfabrik of Augsburg – became his product champion.

Practical experiments soon proved that Diesel's major problem was fuel injection – introducing an exactly metered quantity of fuel to the combustion chamber in a precise fraction of a second while the cylinder pressure stood at some seventy kilogrammes per square centimetre.

Maschinenfabrik stood by Diesel loyally for years while he wrestled with the problem. The company's willingness to support him was remarkable because his original engine idea was only on paper – he hadn't built a working model when they took him on. Their faith was rewarded, for Diesel eventually produced a reliable engine which developed high horsepower with an astonishing thermal efficiency of over twenty-six per cent. Even the most sophisticated steam engines had never bettered fifteen per cent.

Diesel became a millionaire from issuing licences to manufacturers in Europe, Russia and the USA, proving that operating the system instead of fighting it is the best policy. But he was reckless with money, spending it on stylish living and making speculative investments which lost a fortune. Creditors began to hound him. He died at the age of fifty-five, after falling overboard from a cross-Channel ferry in circumstances which strongly suggested suicide. The Rules for Inventors do not cover this kind of accident.

The most striking example of working the system with a patent to gain a product champion was James Watt. Forget those stories about the young James getting his idea from watching a kettle boiling and bubbling on the stove at his parents' home in Greenock. They have nothing to do with Watt's actual career.

It wasn't until he was twenty-nine that he improved Newcomen's beam engine by putting a steam jacket around its cylinder (to keep it hot), and condensing the used steam with a cold water jet in a separate cylinder to create a vacuum. These ideas he patented. His new engine did not sell, because it was badly made and therefore inefficient. Watt had to earn his living surveying canals when the money failed to come in. Ten years later he went into partnership with Matthew Boulton after the pair of them had persuaded Parliament to extend the life of Watt's patent.

The engine took years to come into profit, during which time Boulton subsidized the work from his own manufacturing endeavours in Birmingham. He was famous as a silversmith and maker of ormolu pieces – gilded bronze used to decorate furniture and vases.

Boulton ran a mint whose coins were so good that counterfeiters could not copy them with enough accuracy to defy detection. He was a member of the Lunar Society, the Midlands scientific group which included Erasmus Darwin (grandfather of Charles), Josiah Wedgwood the potter and the painter Joseph Wright of Derby. Boulton was a natural networker who never stopped learning.

There was nothing second-rate about Boulton. His optimistic, outgoing personality and drive complemented Watt's introversion and Scots reticence. He set up a foundry in Birmingham's Soho to make Watt's engines to the highest possible engineering standards. Watt responded with further improvements to the engine. The pair set out to make every penny they could from Watt's patents. This was intensely resented in Cornwall, where mine owners thought that they should not pay licence fees for the economies in fuel given them by Watt's invention. Boulton and Watt won their legal cases against the pirates.

Watt's name lives because he pioneered the conversion of the beam engine into a steam engine. But, without Matthew Boulton, he would have failed and been forgotten.

Rule Eight: Get a manufacturer's backing

The story of Nikola Tesla (1856–1943) highlights the wisdom of this Rule. Born in Croatia and fluent in no fewer than seven languages, Tesla was the epitome of the mysterious Central European inventor. He was tall and leanly handsome, with a piercing eye and an aesthetic and compelling personality. In the late 1880s he burst on the American scene with his new polyphase electrical systems, and amazed the world by pioneering X-rays (before Röntgen), wireless (before Marconi), remote control and robotics. The modern unit of magnetic flux density is named the tesla in his honour.

Tesla achieved his greatest success in defying Rule Six: Don't Become a Dead Hero, which he did by following Rule Seven: Operate the System, and Rule Eight: Get a Manufacturer's Backing. For a time he became a millionaire, courted by the press, for whom he always had some new and miraculous demonstration. When he defied all three Rules simultaneously Tesla fell from heroic status to become a recluse whose sole friend was a pigeon living in New York's Bryant Square.

Tesla arrived in America in 1884, carrying a letter of introduction to the great inventor Thomas Edison. He worked hard and produced an effective arc lamp for street lighting, but was so badly paid that he felt that Edison was exploiting him. Penniless, he left Edison's employ and founded his own company to make alternating-current power systems, generators and motors.

What started as a personal disagreement with Edison became war to the knife. Edison was not only the inventor of direct-current (DC) electrical systems, but the businessman who controlled their use. He owned DC power stations and their distribution networks in New York, London and other European capitals.

Direct current had problems. High current overheated the cables, and DC could not be sent along transmission lines for any distance without major energy losses. A fresh power station was required at one-mile intervals to boost the supply. Tesla's polyphase system did not suffer in this way. It used alternating current (AC) at high voltages, which could be sent for great distances, with no power loss, through light and cheap cables. At the delivery end, Tesla devised transformers which used the power for anything from a giant aluminium smelter to a domestic doorbell.

A lifetime of being right about innovation made it difficult for Edison to admit that Tesla's idea was better. Personal factors came into play, not to mention Edison's enormous investment in DC. He knew that he held the commanding ground and ruthlessly exploited his advantage. Tesla was lined up to become a dead hero.

Cunningly, Edison persuaded the New York state legislature to adopt AC for its new method of executing criminals in the electric chair. Technically, this was sound advice; AC *is* more efficient in electrocution than DC. William Kemmler, the first man to die, expired in conditions of peculiar horror. Smoke rose from his back *before* he was dead, and the execution lasted twenty minutes. Edison then carefully pointed out to the public the deadly nature of AC!

What saved Tesla and allowed him to fight off this propaganda? What let him challenge the well-established DC set-up, and beat it? He had the superior product, there was no doubt about it, but so did Brunel with his broad gauge, and that, after a valiant struggle, went down to defeat.

Part of the answer is that Tesla was fortunate in his timing. If he'd tried to introduce his polyphase system twenty years later than he did, so much money would, by then, have been invested by banks and finance houses in DC power stations and distribution systems that AC would have been brushed aside. Vested interests couldn't have afforded to let it be used. They would have *made* DC work, even if it meant building a power station on every square mile of the earth's surface where electricity was required (and that's what would have been necessary).

The second part of the answer is that, although he probably did not realize it, Tesla operated the system. He found a product champion. His saviour was George Westinghouse, himself the holder of many patents for air-brake systems. They permitted the brakes on every carriage or wagon of a railway train to be applied at the same time. Westinghouse became interested in controlling brakes with electric signals and this brought him in touch with Tesla.

Westinghouse had manufacturing capability; he had money; and he believed in Tesla. The two of them took on the might of Edison so effectively that, in 1895, the world's first AC power station was opened at Niagara Falls. It was a triumph. With his product champion's aid, Tesla beat the rule about challenging an existing system. Niagara Falls sent power over long distances; it fed factories and heavy industry, and lit homes in Buffalo, twenty miles away. AC became the electrical system which powers the world.

These were Tesla's glittering years. He found a way to make high-frequency alternating current from a low-frequency source. We best know his invention as the coil which converts DC from a car battery into AC to make the engine fire when it starts up.

Tesla lit his laboratory with the forerunner of fluorescent lighting. He took X-ray pictures of his own skull and the bones of his hands before Röntgen published his own researches.

Research gave Tesla the idea for wireless, whose working he demonstrated before Marconi became famous. At Madison Square Garden, in 1893, he showed a model boat which he sent this way and that with wireless signals. Tesla's radio control is now used all the way from crop-spraying aircraft to guided missiles. But when he first demonstrated it Victoria was on the British throne and the principal means of transport were steam trains and horses.

The world lay at Tesla's feet. He responded by throwing his advantages away. He ignored the need to operate the system. In 1899 he moved to Colorado Springs to test his idea that it was possible to send power anywhere around the globe *without* the use of wires. He constructed a laboratory with the largest coil he'd ever built, fifty-two feet in diameter. It discharged a fantastic ten million volts. Photographs he published of himself, seated among electrical streamers, yards long, remain riveting images, reminiscent of Frankenstein engaged in the search for life.

Tesla financed the lab himself. He was always careless about money and acted as if the income from his AC system patents would continue for ever. It didn't. And now he committed an act of folly. J.P. Morgan, the financier, was generous to him personally with gifts of money. Tesla let it be understood that Morgan was backing him in the construction of his World Telegraphy station on Long Island.

The station was to be an early version of Broadcasting House, with its own programmes transmitted over Tesla's radio system. The wireless side of the station was to link the world's telephone exchanges, provide signals to regulate clocks, establish a universal marine navigational system and allow reproduction, anywhere in the world, of pictures and drawings. In effect, Tesla was offering the equivalent of what is now provided through satellites: international STD, broadcasts of Universal Time, global positioning information for sea and land navigation, and fax transmissions!

Results were curiously slow in arriving and when Tesla started trying to find extra funds to carry on with building the station, word got out that Morgan had never invested a penny in the scheme. Tesla had not only failed to operate the system – he had completely alienated it. An awful warning. He never found another backer.

Tesla later confessed that, as well as starting wireless transmissions, he had planned to send power all round the world without wires. By shooting at two targets, Tesla missed both. There's little doubt that he was suffering from delusions of infallibility – understandably so, in the light of his record. He also broke Rule Two: Don't Be Before Your Time. The tragedy is that he was technically correct – a version of his World Telegraphy station could certainly have transmitted messages – but Tesla was fatally ahead of the imagination of his contemporaries.

CHAPTER EIGHTEEN

The Personal Equation

'O! I have lost my reputation. I have lost the immortal part of myself!' Shakespeare, as usual, has it right in *Othello*. Innovators care about money, of course they do; don't let one of them ever persuade you otherwise. But they also care about reputation – 'the immortal part' of themselves.

To become a successful inventor is to make a mark: 'I was here.' In the same way that a medieval mason put his sign on the stonework he added to a cathedral, it's a tiny piece of immortality. The cruellest of disappointments is to gain no credit when credit is due, to conquer a host of problems and then be ignored or forgotten. The cause may be the personality of the innovator, bad timing, failure to publish, or even a matter of nationality.

Jacques Charles (1746–1823) was a brilliant publicist who gave lectures illustrated by practical experiments on the behaviour of gases. He was deeply influenced by the researches of Benjamin Franklin into lightning and the methods of preparing gases developed by Dr Joseph Priestly. On 1 December 1783, after a number of unmanned flights, he took off from the Tuileries Gardens in Paris in a balloon filled with hydrogen. It reached the amazing altitude of 3000 metres during a flight which ended over forty kilometres away.

Was Charles accorded a world first? No. His timing was exquisitely wrong: the Montgolfier brothers had flown in a hot-air balloon ten days previously. However, he reaped one unexpected reward from his work. When a mob invaded the Tuileries at the outbreak of the French Revolution nine years later, Charles saved himself from being torn to pieces by reminding the blood-crazed crowd of his ballooning exploit. It had so impressed the working people of Paris that his life was spared.

He continued researching into gases to discover that their

volume is proportional to absolute temperature at constant pressure. His experiments showed that a gas's volume increases by 1/267th for every degree rise in temperature. Charles had discovered the existence of absolute zero, fifty years before Lord Kelvin. He tried to throw away this second chance of fame by not publishing his results. His discovery was only named Charles' Law because another scientist published it for him. Some discoverers almost deserve being forgotten.

The inventor's nationality can be important. We tend to be chauvinistic about priority in discovery and attribute it to inhabitants of our own homeland. If that land is a powerful economic force, the rest of mankind tends to let its claim to priority be rammed down its throat. Inventors from smaller nations become forgotten outside their own national boundaries.

Consider the debate about the pioneer of mass production. Credit (if credit it be) usually goes to Richard Arkwright for starting large-scale mechanization with division of labour between machines and the workforce.

As the eighteenth century progressed, manufacturing in the north of England was increasingly based on machines powered by water-wheels. Kay's flying shuttle of 1733, Arkwright's frame of 1769, Hargreaves's spinning jenny of 1770 and Crompton's mule of 1779 were all driven by water power.

Arkwright's frame revolutionized spinning by drawing out rovings of cotton and twisting them as they passed over two sets of rollers, the second travelling quicker than the first. Extra capacity created by the frame's speed of operation created a bottleneck in the supply of carded rovings to the frame. Arkwright devised a faster carding machine. It combed the fibres of raw cotton to make them lie parallel and then pulled them into further alignment.

But neither of these innovations was his crucial contribution to industry. This came when he installed machines at his mill in Cromford, Derbyshire, in 1771 and took on hands to work them. By 1775 he had a network of factories around the country, carrying out all cotton processes from carding and drawing to spinning. Seven years later he employed no fewer than 5000 workers, some as far away as Scotland. The factory system had begun.

Another claimant to the title of introducer of factory production is the fascinating Frenchman, Jacques de Vaucanson (1709–1782). Suitably enough, he began his life building automata, based on his studies in anatomy, music and mechanics. His *ne plus ultra* was a mechanical duck which swam, quacked, flapped its wings, ate grain and then excreted an appropriate product. It was shown at the court of Louis XV to great applause – the French aristocracy had a simple sense of humour.

De Vaucanson's court connections and mechanical skills helped him to become Inspector of Silk Manufactures. He was a brilliant toolmaker, designing the first screw-cutting lathe and an improved drill. In 1756 he set up a silk factory at Aubenas, near Lyons, which predated Arkwright's mills by nearly 20 years.

He made a mechanical loom for weaving figured silks (materials with patterns woven into them). It was reassembled in 1804 to provide the basis for the Jacquard process, which has been superseded by computer controlled machinery only recently.

Here, industrial historians in the UK and France part company. The British claim there is no evidence that the Frenchman's loom worked properly – it had to be rescued from oblivion and redesigned before it could become the Jacquard loom. Furthermore, the silk factory was a dead end, an inspired one-off; it created no precedent which others followed.

Now, it doesn't matter who is right or wrong in this debate. The fact is that De Vaucanson's name is linked to the factory system only in France; Britain's industrial dominance during the eighteenth century ensured that Arkwright wins the credit everywhere else. (Be honest: if you are neither French nor a mechanical engineer, have you ever *heard* of De Vaucanson?)

Moreover, the Frenchman's claim can be set aside because there is a much earlier one to be made for Christopher Polhem (1661–1751), a Swedish mining expert.

Mining was the major export industry in Sweden, successful sales of iron were needed to keep the trade balance favourable, and the Falu Mine was the largest employer in the country. At the mine, Polhem gained a flying start as an innovator by constructing several water-driven machines which lifted ore out of the shafts. In 1697 he came up with a plan by which all its pumps and hoists would be powered by enormous water-wheels far away from the mine itself. He gained the ear of the king, Charles XII, and together they planned to improve the naval base at Karlskrona and cut a great canal across the country.

The king raised Polhem to the nobility and gave him a well-paid appointment. Ah! those were the days. Maybe, if similar honours were handed out today, the English class structure would flex to give engineers and innovators the respect they deserve and encourage more people to take up invention as a career.

Polhem's great breakthrough came when he set up Stiernsunds Bruk in southern Dalarna in 1700, a water-powered factory for the manufacture of tools. Polhem wanted to mass produce metal implements. He recognized the economic sense of the division of labour and was an ardent advocate of replacing muscle power with machines. Stiernsunds Bruk mostly made smaller iron articles such as scissors, mirrors, locks and clocks, with all processes as mechanized as possible. The machines shaped hammers for plates, and cut files and cog wheels; sheet metal was made in a roller mill which had no equal in its day.

But Polhem's factory was not a winner. The water-powered devices available to him were not as reliable as Arkwright's. The Englishman was working in a flourishing economy, the Swede in a country that had been depopulated and impoverished by the European wars of Gustavus Adolphus and Charles XII.

Polhem is known in his own country as the Swedish Daedalus, the personification of ingenuity. But, like De Vaucanson, he remains largely unknown outside his homeland. Sweden was not important enough in the years following his innovation for the fame of his work to spread.

I have indicated that, when it comes to assigning credit for an innovation, the vital test is whether it reached the stage of a working model and, then, the market-place. Theory that leads nowhere in practice, like Leonardo da Vinci's aircraft designs, doesn't count.

Arkwright will always be credited with starting mass production, because that is what he did. Polhem and De Vaucanson's efforts faltered and died, but there is a clear line of descent from the mill at Cromford to the robot-assembly factories of the present day.

Rule Nine: Success depends on the inventor

The cautionary story of Richard Trevithick illustrates this Rule perfectly. He was improvident, rash and generous to a fault. His amazing life could not be invented, although he was prodigal with invention itself. If reputation can be destroyed by the personality of the inventor, reckless Dick Trevithick is the example.

Richard Trevithick (1771–1833) was a giant of a man, six feet two inches tall and broad and burly to match, known as 'Cap'n Dick' because 'captain' was the Cornish word for a mine manager. He was bursting with charm and energy. When an enterprise folded under him (as they nearly all did), he always bounced back undefeated.

He grew up during the Cornish mine owners' struggle against Boulton and Watt. One of their hopes for circumventing Watt's patent was to invent a steam engine which avoided using Watt's condensing cylinder. Trevithick proposed to use high-pressure steam (Watt's engine was low-pressure) to increase the engine's power and then vent the steam to atmosphere, thus eliminating the condensing cylinder. The idea was highly successful – Watt sourly commented that Trevithick ought to be hanged for thinking of it – and Trevithick's fame spread from Cornwall to the mines of South Wales.

His next move was one of blinding simplicity. He took his high-pressure engine, mounted it on wheels, and drove the wheels by the action of the engine. The result was a road locomotive. The machine, known locally as 'Dick's Firedragon', was tested on the steep hills around Camborne on Christmas Eve, 1801, giving rise to the song 'Going Up Camborne Hill, Coming Down'.

It did not last long. On 28 December the locomotive came off the road after hitting a gully. The merry crew on board pushed it into a shelter before adjourning to a nearby pub. They forgot to draw the fire and, while they were enjoying draughts of post-Christmas ale, the engine burnt down, taking the shelter with it.

Trevithick had destroyed not only his locomotive but the world's first garage.

Unperturbed, in 1804 he became involved in a 500-guinea bet, made by a Welsh industrialist, that he could transport, by steam power, ten tons of iron on a tramway from Penydaren ironworks to the canal ten miles away at Abercynon.

Trevithick's machine – now a real railway locomotive – was triumphantly successful. But it made few journeys beyond the wager-winner because the brittle tramway rails cracked beneath the weight of the train.

Another Trevithick locomotive was built at Gateshead in 1806. George Stephenson, not yet a railway builder, saw it there and talked to Trevithick; there's little doubt that it stimulated Stephenson in his own railway career.

Somehow, Cap'n Dick was unable to make money from his engines. He made a final attempt to exploit them when he built *Catch Me Who Can* in 1808. It ran on a circular track near what is now Euston Station and was intended to prove that it could move quicker and further than a horse. Rides cost a shilling each, equivalent to about £2.50 nowadays – quite a price. But the chattering classes of London ignored the locomotive, the cast iron rails broke under its weight, and the venture failed.

Trevithick gave up work on railways for ever. He tried a series of abortive schemes.

- He attempted to drive a pedestrian tunnel under the Thames, another first. He was within sixty yards of success when the proprietors withdrew their support from him because of delays in the tunnelling. The tunnel would have been another first.

- He took on a partner, who defrauded him over a plan to make a steam tug and floating crane and left him bankrupt.

- He raised a wreck off Margate, using empty iron tanks to provide buoyancy for the submerged vessel. After the ship had been lifted, there was an argument with its owners about payment. Trevithick, careless of legal details, had not agreed a firm contract. Incandescent with rage at what he saw as fraud, he cut the ropes and the wreck sank once more to the bottom of the sea.

- He built a 'recoil' engine, very like the turbine invented by Hero of Alexandria. Its arms, twenty-four feet in diameter, spurted high-pressure steam from nozzles at their tips. Trevithick planned to use the engine to drive a ship, but it disappeared mysteriously from history.

There were successes as well: the Cornish high-pressure engine in 1812 and a steam-powered corn threshing machine in the same year. But his greatest mis-adventure was to come.

The rich silver mines at Cerro de Pasco in Peru had flooded, making them impossible to work. Trevithick supplied four pumping engines which he shipped out with Cornish workmen to supervise their running. Reports came back that the engines were not working properly, so Trevithick decided to go to Peru himself. He did not return home for eleven years.

There's no space here to chronicle in full his adventures in South America. In brief, his engines were smashed and thrown down the mine shaft by anti-government rebels. He was forcibly impressed into the army of Simón Bolívar and, after escaping to become a miner, was robbed of his tools and thousands of pounds worth of silver ore. He made his fortune again with a ship-salvage operation, and lost the lot on a pearl-fishing venture in Panama.

Ever optimistic, the Cornishman was convinced that his

fortune was to be made exploiting the mineral wealth of South America. He met a Scot called James Gerard who told him there were riches beyond imagination waiting for them in Costa Rica. After four years, they decided to go back to England and raise money to work their mineral claims – the exact reverse of what Trevithick had in mind when he set out to make his fortune.

To shorten the journey home, Trevithick decided in a casual sort of way to walk across the isthmus to the Caribbean – a trip which first involved crossing the Cordilleras and then plunged into almost impenetrable jungle. Incredible though it seems, Trevithick and Gerard took with them two boys on their way to school in Highgate!

It was a journey of immense hardship and privation. They were nearly drowned in the River Serapique and lost their equipment; one of the natives accompanying them was swept away and never seen again. They lived on monkeys and wild fruit; their clothes were torn to rags. Trevithick was half-drowned a second time and nearly eaten by an alligator. Eventually, they (schoolboys included) reached the coast and made their way to Cartagena.

Here occurred the most incredible incident of all. Trevithick somehow managed, despite his emaciated appearance and ragged clothes, to gain entry to an hotel, where he found a diffident young Englishman was also staying.

'Aren't you Captain Trevithick?' asked the young man, introducing himself to the tatterdemallion figure. Trevithick peered at him. 'Why!' he exclaimed. 'Is that you, Bobby?'

The stranger was Robert Stephenson, the son of railway-building George. The Cornishman had known Stephenson junior twenty years before as a small boy while his locomotive was being built on Tyneside. Robert was in Colombia on a mining mission whose lack of results matched those of Trevithick himself. But *he* was not hungry and penniless. He generously gave the older man £50 to pay his passage home and the devil-may-care Cap'n Dick returned home, poorer than when he left eleven years before.

This spurs me to ask: why is George Stephenson known as the Father of Railways while Richard Trevithick, their true inventor, is almost forgotten? It is largely Trevithick's own fault. He was so fertile in innovation, of such a quicksilver mind, that he found it difficult to concentrate on making one good idea succeed. If he had persevered with his locomotives and found a product champion among the many industrialists he knew, the story might have been different.

Stephenson, on the other hand, was intensely practical. He improved the locomotive itself by using steam blast, which increased the draught through the firebox and made the locomotive more efficient. He introduced flanged wheels, ending a long debate about whether they should be on the wheels or on the rails (a point not understood many years later by Alfred, Lord Tennyson, who referred to great wheels spinning for ever, 'down the ringing grooves of change'). Stephenson also worked out the theory of level lines with viaducts and cuttings, and built the first locomotive with boiler tubes, which vastly increased the heat-transfer efficiency of the machine.

Few of these innovations were his own, but Stephenson was a genius at making other men's ideas work. He understood that with railways he was not dealing with just one invention, but a *system*. Trevithick should have shared this insight and acted on it. In particular, he should have noticed that two of his railway failures were caused not by failure of the locomotives, but of the track; one part of the system was letting down the rest. Ignoring this, he went rushing off to tackle new endeavours and so gave up his greatest invention.

Although he was lion-hearted and indomitable to the end, Trevithick died so poor that his fellow workmen had to raise a subscription to pay for his funeral. He lies in an unknown grave at Dartford in Kent, far from his beloved Cornwall.

Rule Ten: Persistence leads to success

There are many cases where persistence eventually won the day. Not bull-headed plugging away with an idea that was incapable of working, but steady elimination of problem after problem. Diesel, with his engine, was an example. So was Alastair Pilkington with float glass. They were both able to continue their work because they were backed by product champions, although it's not often possible for independent innovators with limited private resources.

The story goes that, in the 1950s, Pilkington was washing some dirty dishes when the sight of a plate floating on the water suggested to him the idea of float glass. There was a demand for optically correct glass in mirrors and shop windows and road vehicles. In those days, it could be produced only by expensive grinding and polishing processes; the market wanted a cheaper but equal product without the distortions of the mass-produced sheet glass then available.

Pilkington's concept was to float a continuous strip of glass out of the furnace on to molten tin. The surface of the tin was optically flat and, when the glass travelled at the right speed over it, it would assume the same shape as the tin. After a further interval, the glass would cool off so much that it could be removed from the tin without suffering any mechanical damage. The result was a bright, self-polished glass of uniform thickness.

It took Pilkington seven years to get the process right. The cost was phenomenal and it must have taken the board of Pilkington Brothers (that they had the same name as the inventor was a coincidence) all its nerve to continue. But continue they did, with the result that float glass made the company's fortune. It is used not only in cars and mirrors but also in spectacles, aircraft and defence applications. Persistence paid off.

We're so far away from buttons and bows that we've forgotten what an incredible nuisance they were to wear, what a fiddly business it was to do them up and, more so, to undo them – especially if you were in a hurry. The common-or-garden zip fastener that replaced them took decades to bring to perfection. The earliest version was patented in 1891 by Whitcomb Judson, an American mechanical engineer.

A row of hooks and eyes locked together when you pulled on a slide. This primitive zip was sold as an easy-to-use substitute for rows of couplings on shoes and high boots, which required a buttonhook to undo them. It didn't sell well because it had a disconcerting tendency to open up on its own – usually at inopportune moments. In 1900 Judson produced a version called C-curity which extended the zip's use to clothing by attaching it to fabric. C-curity was the wrong name; secure it wasn't.

Judson struggled along until, in 1906, one of his employees, a young engineer by name of Gideon Sundback (which sounds straight out of Moby Dick), designed a fastener with interlocking metal teeth, latched together by a slider. Each tooth was a little hook which gripped an eye under an opposing hook on the other half of the device. The slider spread out the teeth as it was pulled, leaving them engaged in its wake.

But the invention's struggles were not yet over. It required manufacturing to a high standard of precision, but not by an expensive process which made it too costly for mass use (just as in the case of float glass). It was not until 1913 that Sundback produced a machine to stamp out the teeth and crimp them to tape in their millions. It had taken twenty-two years for the idea to become commercially feasible (and it wasn't until 1926 – thirty-five years after the first patent – that it was christened the zip fastener). Persistence won the day.

You'll remember that I set out to discover what lessons could be learned from the successes and failures of past inventors. As a result, we have now accumulated my second Ten Rules for Inventors, derived from history. How do they match the original Ten Rules, drawn from my own experience and that of many colleagues in the business of innovation? Do they support my arguments, or undermine them? If I am right, they ought to match in most respects because, as I have pointed out, the process and principles of innovation do not change with the passage of time.

Here then are the twin sets of Rules, set out in numbered pairs, with brief notes underneath each.

Bob Symes' Combined Rules of Invention

Rule One
Identify the problem.
Invent for tomorrow.

Comment: if you identify a problem, you find one that has not been solved today; you are inventing for tomorrow.

Rule Two
Meet a need.
Don't be before your time.

Comment: if you are before your time, it may be that you are ahead of the necessary technology to achieve your end, and it may be that society does not require your innovation. Either way, you cannot meet a need.

Rule Three
Keep on learning.
Experience is never wasted.

Comment: these are the same.

Rule Four
Check for originality.
Join a network.

Comment: not a match, but I didn't expect 100 per cent correspondence of ideas. However, checking for originality is a necessity under patent law, and joining a network (of colleagues, information providers, fellow innovators etc) is enlightened self-interest.

Rule Five
Build a working model.
Prove your invention works.

Comment: the same.

Rule Six
Don't attack established interests.
Don't become a dead hero.

Comment: there's no more assured way to become a dead hero than by taking on giant companies and putting their interests under threat.

Rule Seven
Learn the patent system.
Operate the system.

Comment: if you don't learn the system, you cannot operate it to your advantage.

Rule Eight
Find a product champion.
Get a manufacturer's backing.

Comment: the same.

Rule Nine
Sell yourself as well as the invention.
Success depends on the inventor.

Comment: the same.

Rule Ten
Persevere.
Persistence leads to success.

Comment: the same.

A score of nine out of ten, with the single exception having both of its propositions self-evidently true. You don't have to take my word for it that the first Ten Rules work. They are proved by experience. I rest my case.

CHAPTER NINETEEN

Where to Find Help

'I was not aware that any organizations existed that were able to help me. A guidebook listing such help would have been useful.'

'I have had no help from government organizations such as the DTI because I was never told how to go about seeing people from organizations etc.'

Despite the slight note of self-pity in these comments, it's true what the inventors say. Potentially valuable new ideas wither and die because their owners don't know how to go about developing and selling them. The only satisfactory current guide for innovators that I know of is published by the Design Council; but this guide is itself under threat, because of changes in government policy.

However clever your invention may be, a time can come when things go wrong, when you get stuck, or when you are so baffled by continual rejection that you don't know what to do next. I have felt that way myself at times.

There is no formula to persuade other people that your idea is good and necessary. No magic wand exists to extract money from hard-faced investors against their better judgement. But there are many organizations which can give assistance in your difficulties. I don't want to clutter up my account of who they are and what they do, so in Chapter Twenty you'll find their names, addresses and telephone numbers, as well as those of many other organizations.

But, before begging for help in making or promoting your product, the crucial thing is to try solving problems for yourself. Don't go off at half-cock. Remember Rule Nine: Sell yourself as well as the invention.

You can actually compound your difficulties if you come across as unprofessional. Put yourself in the other person's shoes. Would you appreciate being asked for help by questioners who haven't taken the trouble to do their basic homework?

Patents

For instance, don't go asking questions about patents before getting in touch with the Patent Office's Information Service. This is the first port of call for anyone seeking to discover how the patent system works. The Office supplies a superb range of *free* literature with facts about patents, trade and service marks, copyright and design registration. No budding inventor should be without the Office's five booklets:

- *What is Intellectual Property*
- *Patent Protection*
- *How To Prepare a UK Patent Application*
- *Trade and Service Marks*
- *Patent Search Service*

Self-help

Next, consider joining the leading organization for private inventors. I have to own an interest here: I am its President and former Chairman. But it has the great advantage of having no axes to grind, no interests which might work against yours without you knowing it, no shareholders or commercial investors who hope to make money themselves from *your* ideas.

The Institute of Patentees and Inventors (IPI) is the largest association of inventors in the western world, with 1500 individual members, as well as companies belonging as corporate subscribers. It is non-profit making. The Institute is a body that can help at a certain stage in your innovating career. This period is best described as lying between the time when you produce your working model (if it really works!) and when you are selling well enough to need no more assistance.

The IPI grew from small beginnings in 1919, when five pioneers met in the chambers of a barrister in London's Temple to plan a 'Society for the protection of Patentees and to establish an extension to the life of Patents'. It puts out a quarterly magazine, *Future and the Inventor*, which covers stories of interest to private innovators.

For a small fee, members (and non-members) can promote their work and offer it for licensing through the widely circulated *New Patents Bulletin*, which appears six times a year. Companies showing interest are then introduced by the Institute to the innovators and their inventions.

The Institute takes space at exhibitions such as those organized by The Engineering Council. It acts as long-stop for many inquiries initially made by members of the public to bodies like the Department of Trade and Industry and the British Technology Group.

Although the IPI doesn't finance inventions itself, it has a network of experts, contacts and business relationships built up over the years which can assist members with their problems. I can almost see the gleam in the eyes of some innovators at the faintly implied suggestion that the Institute can help to find finance for ideas. Believe me, there is *no* such promise.

But the IPI is one of the networks to join if you are to succeed. Who can tell which one will eventually pay off? And it does confer a comforting feeling that you are not alone in the world, with every man's hand against you.

Note, however, that the Institute is *not* a patenting body. It does not process patent applications – that remains the inventor's responsibility. Nor can it give extended advice to people who ring up for free information. Please find out what the IPI has to offer, but do the decent thing: become a member before seeking counselling about problems.

Members of the IPI have included Frederick Mackenzie, the deviser of Letraset, Ron Hickman, who invented the Workmate, and Sir Barnes Wallis, designer of the R100 airship and the bouncing bomb.

General Assistance

In the course of developing ideas you must rely on the aid of outsiders ranging from specialist suppliers and testing bodies, to manufacturers, designers, and marketing experts. In consulting them, you must do the very thing that you don't like doing: reveal the idea.

And then there's always the danger that the search for help will be rebuffed. Commercial people may see you as a dangerous challenge to their existing products because they did not come up with the notion themselves. Remember NIH?

Worse, if the innovation is not properly protected, they can adopt it as their own and leave you to whistle. Even if it is hedged about with legal safeguards, they may decide the money to be made from stealing it is so great, and the innovator so defenceless, that theft is a chance worth taking.

On the other hand, you do need assistance. How is this problem to be resolved? One answer is to go through official bodies which promise to observe confidentiality.

THE DESIGN COUNCIL
There are government plans to remove nearly all the Design Council's innovation activities and hand them over to other organizations. These plans were announced in such a rush that the 'other organizations' remain unidentified. The Council continues (temporarily) to run its services so I describe them as they exist. But change is on its way.

I can't speak too highly of the work that the Council has done through its Innovation Service, work which is conducted in conditions of commercial secrecy. Innovation NoticeBoard is the most notable part of the service. It puts new products and ideas in touch with manufacturers.

Case history
A partnership between a development agency and research scientists produced a construction called Space Frame, a load-bearing roof constructed entirely from glass-reinforced plastic. The partnership approached NoticeBoard, hoping to find an industrial company which could make and sell the idea commercially. Space Frame went on display at the Design Council.

Glynwed International plc, a construction and engineering company employing over 12,000 people, saw the exhibit and eventually signed a licence agreement with the Space Frame partnership. Space Frame's design, with threaded plastic joints supporting the roof structure, fitted well with Glynwed's work in its plastics division.

How does the NoticeBoard work? Fill in a form, spelling out the details of the innovation. A panel of experts then assesses it for commercial potential. This is no easy run – you have to prove your worth. But if you succeed, the product is listed on the NoticeBoard database, which tries to match new products with interested manufacturers. Over 800 companies are on the database. They cover engineering, electronics, and consumer goods and may be able to help in many areas – product development, manufacture, finance, distribution and marketing.

Case history
TrimEasy is a hand-held, battery-operated wallpaper trimmer which uses two pins moving in and out at high speed to perforate the paper, giving a clean finish with no torn edges. Inventor Alec Janaway was advised by Innovation NoticeBoard staff to

consider making TrimEasy himself, to prove its market possibilities to manufacturers.

The Design Council's Design Advisory Service also gave him advice on materials, product testing and potential manufacturers. As a result, TrimEasy is now widely available. The Council quotes Alex Janaway as believing that guidance on the innovation process is one of NoticeBoard's greatest strengths. The NoticeBoard service costs nothing because it is funded by the Department of Trade and Industry.

The Design Council's Contacts Directory is also worth a note. Do get hold of a copy – if it is still available. In addition to a comprehensive list of useful organizations, it has three pages of information on the innovator's Eldorado: sources of venture capital. Don't become too excited by this. The organizations involved operate to the most rigorous standards: the innovation must be outstandingly good and useful, and inventors have to prove that they can make it in business.

The best advice I can give is to stay away from venture capitalists until everything you need – working model, costings, production protection, and business plan – is thoroughly worked out. Find a few pedantic professional friends (for some reason, accountants or lawyers spring to mind) and get them to subject you to the third degree. Tell them to quiz you about things that can go wrong: commercial and marketing prospects, cash flow, alternative suppliers – all the hazards that have brought down many likely entrepreneurs.

When you're fluent in answering questions (if you don't become fluent, you are in trouble), *then* approach the venture capital company with your carefully thought out presentation. I wish you the best of luck.

The Department of Trade and Industry (DTI)

The DTI runs a Freephone Innovation Enquiry number which will answer questions about the Department's innovation schemes. The first of these is SMART: *S*mall firms *M*erit *A*ward for *R*esearch and *T*echnology. This scheme offers government money and is open to small firms in the UK employing fewer than fifty people, and to individuals, as long as they are starting a new business to exploit an invention. SMART money covers the whole range of technology but communications, instrumentation, biotechnology and computer-linked ideas are especially welcomed.

The rewards are good, but SMART is tough. You'll be assessed for innovation and originality, but also for serious commercial points, including sales potential, a workable business plan and capability to exploit the idea successfully. You'll be competing against 1500 other projects. Don't bother to apply with something as simple as an improved wheelbarrow (though a product that is as simple and innovative in concept as, say, the Sony Walkman is OK. Getting the electronics right was the only difficult bit!)

SMART awards come in two stages. Stage 1 offers up to £45,000 towards the costs of feasibility studies in one year. Winners qualify for Stage 2, which can give up to £60,000 to develop a prototype over a further year.

SPUR (*S*upport for *P*roducts *U*nder *R*esearch) is for companies with up to 250 employees. The award helps small firms to develop new processes and products which involve a significant technological advance. Grants of up to £150,000 a year are available.

The SMART and SPUR schemes are not fixed for ever. They change all the time, and the government could kill them off tomorrow. Ring DTI Innovation Enquiry 0800-442001 for up-to-date information.

Competitions and Awards

Most schemes are for younger people – some of school age, some older. This is entirely right, for innovation often comes from the new generation and needs encouraging. Yet young people with new ideas are those with the least money to back them up, because they have not had the time to accumulate capital. The awards listed here can help to solve the problem.

The Prince of Wales Award for Innovation

This is the sexiest contest – partly because of Prince Charles's backing, partly because finalists are publicized by appearing in a special edition of the BBC's *Tomorrow's World*.

The way the award is given has changed. Under the former system, six finalists for the 'innovation' stage were chosen each year and over the next two years their progress in developing their product and making it was closely monitored. Finally, an outright winner of the 'production' stage was selected. Entries divided neatly into one third from private individuals, one third from small firms or partnerships, and the rest from big companies and institutions.

Now the scheme has been altered to give more emphasis to successful selling. It is more hard-nosed. Winners are picked year by year, and there are two categories of award:

Innovation: ideas which have good commercial potential but which are not yet proven in the market-place.

Commercialization: innovations which have already demonstrated commercial success.

The awards are run by Business in the Community, 400 businesses working with government, the unions, and voluntary and community organizations to promote investment and involvement in local communities and inner cities.

SHELL TECHNOLOGY ENTERPRISE PROGRAMME (STEP)

Strictly speaking, STEP is not an award scheme for innovation. It is a sponsored plan which helps young people learn to innovate – the only one of its kind, as far as I know. You'll have noticed that I am keen on youngsters training themselves in innovative thinking, so I feel that it's legitimate to include STEP here. Every year STEP finds undergraduates short-term placements in small or medium-size businesses, where they work on projects of practical value.

Interestingly, newer-established universities almost swept the board in the last award list. Apart from a lone entrant from Sheffield University (who turned out to be overall winner), awards went to candidates from De Montfort, Nottingham Trent, Warwick, Robert Gordon and Ulster Universities.

Case history

Nicholas Latham, a Sheffield University undergraduate, invented a reporting system which let the St Alban's Rubber Company in Durham work out how well it uses its raw material.

During an eight-week summer stint, the twenty-year-old electronic-engineering student analysed working methods on St Alban's shop-floor to measure their efficiency. He found two ways of using materials better, one of which was taken up by the company.

John Meredith, St Alban's production director, commented: 'The enthusiasm, commitment and clear thinking Nicholas demonstrated has had a profound effect on all our employees and management, causing people to question why they were doing things and whether a better way was possible.'

Case history

A mathematics and computing student from the Robert Gordon University, Hazel Gordon, computerized the job estimating system for Salamis Ltd, a fabric maintenance company. Salamis's commercial manager, John Innes, noted: 'Hazel has achieved more in two months than the company has in the past two years.'

Case history
Hilary Grist, a law and marketing student from De Montfort University, carried out a market-research study into the foam-lined grey-board market for S.H. Fiske Ltd, to produce a marketing plan.

'Hilary has brought a breath of fresh air to our company,' managing director Richard Jarrat said. 'She has created from scratch a sure-fire marketing plan, ready to go, which cannot fail to be profitable. But better still, she has changed completely the way we look at our products, markets and what we can do with them.'

All this is further proof of my suggestion that young people are free of preconceptions, that they approach problems with a fresh eye, and that they can see what more experienced people miss. And notice how well the girls are doing!

Whatever age you are, try to behave the same way: *think like a child*.

YOUNG ENGINEERS FOR BRITAIN
This is an annual contest for young people, either in full-time education or in industry. Entry is through school or college or company. It's essentially a regional competition, held in a dozen centres throughout Britain.

Case history
I was moderator of the judges when Lucy Porter from Bath High School won the title of Young Engineer for Britain. It was satisfying that they chose her swing exerciser for handicapped children as winner, not only because of the simplicity of the idea but because Lucy taught herself engineering. She came up with her invention after helping to care for children with special needs at the Royal United Hospital in Bath. One little girl couldn't use her legs, so was unable to play on swings or other equipment. Lucy decided to design something for children like her.

Although she was an arts student, Lucy bought textbooks and made engineering drawings in her own time. She produced a first-rate working model of her swing, which has a moulded seat with a ledge that prevents the occupant from sliding out. The swing itself is worked by a pull-handle, rather like an old-fashioned fairground swing-boat (see photo section).

It was the second year running that a girl won the competition, proving once again that innovation is not simply a man's game.

Any type of engineering is welcome in Young Engineers for Britain. Marks are given for originality, skill, the right use of scientific principles and presentation – that means good talking and good written material – and good working models! Marketability, usefulness and *meeting a need* also count.

The contest is divided into four classes by age. There are class prizes of cash, plus tours of famous companies such as Shell, GEC and the Rover Group. The overall winner gets a trophy and £1000, and his/her school wins another trophy and £1500 to buy engineering equipment.

A cornucopia of special prizes includes goodies such as sponsorship for degree courses in engineering. Entrants to the national final have been offered help with the commercial development of their projects under a scheme funded by the Comino Foundation.

The best project by a girls' team wins the WISE award (Women Into Science and Engineering). I begin to wonder if girls need this positive discrimination when they are making such hay in the contest for the premier award!

Rotary Young Inventor of the Year

Following the Young Engineers for Britain event, Lucy Porter (see above) also won the £10,000 prize awarded in the Rotary International/Plus Plan Young Inventor of the Year contest. Competition was fierce. The finalists were drawn from an original entry list 5000 strong which included a no-hands page-turner for musicians, a yacht immobilizer, a walking stick with a built-in torch and alarm, a false start detector for swimming races and an autocue for copy-typists.

Lucy beat twenty-seven other finalists, becoming the first girl to win the title. Her story goes to support my theory: that anyone can become an inventor if they have enough determination, and anyone can train themselves.

Young Engineers National Awards

Young Engineers calls itself 'the dynamic, modern organization for young people interested in inventing – and having fun at the same time'.

It's mainly educational set-up (don't go away, this is a good story) run by the Standing Conference on Schools' Science and Technology, which operates through clubs based in colleges and schools (including primary schools). Young Engineers' approach is to show young people how much they can gain from inventing, largely as a group activity, but also as individuals. Some of the work is really impressive – thanks, of course, to the dedicated work of teachers involved.

Case history
Project Volta at St Richard's School in Bexhill, East Sussex, set out to make a serious attempt on the world land speed record for light electric cars. It aimed to beat the 100.242 kilometres an hour set in 1981 by Jens Knoblock in Germany.

Two boys and two girls formed the main design team with the guidance of Design Technology teacher Peter Fairhurst. Industrial back-up for the project was immense. Lotus Engineering, DETA batteries, ATS, Michelin tyres, MIRA, Seeboard, the National Westminster Bank and the Cookson Technology centre at Oxford were all involved in supplying materials or money to support Project Volta.

The attempt on the record was made at Greenham Common airfield in Berkshire, with Lotus test driver Rudy Thomann at the wheel. He described the vehicle as very noisy inside, but with better acceleration than a racing car.

Project Volta clocked up 106.74 kilometres an hour, and, subject to official ratification, St Richard's Young Engineers team beat the world record (see photo section).

This is where the future lies, in the hands of schools. The young people involved aren't learning only engineering. They are involved with industry, with project management, with design economics, with sales – with every skill that a successful entrepreneur must master.

And there are prizes, at the Young Engineers National Awards. The club of the year wins £1000 cash and the European Experience, a visit to a company internationally known in Britain and Continental Europe.

Girls are encouraged with a special award for an all-girls club. Seven clubs are invited to send three members and their teachers on a week-long study tour, there are a host of specially arranged visits to edge-of-technology companies, and the prizes in total rack up to more than £60,000.

CHAPTER TWENTY

Directory of Contacts

Organizations change. They come and go – especially those designed to help inventors! Lamentable losses in recent years have included the Association of Innovation Management, a non-profit-making trust which ran seminars on inventing and offered panels of experts who could be approached for advice, and the Toshiba Year of Invention, the biggest and best of the innovation competitions. And with one gesture, the government has stripped the Design Council of the services which it offered to innovators. Any scheme under government control can have its throat cut without notice.

Telephone numbers change. From 16 April 1995, all the UK telephone numbers I give here must have a 1 inserted after the first zero e.g. 071 (central London) becomes 0171; 0633 (Newport) becomes 01633. The international access code alters from 010 to 00.

So, a word of warning. The directory is up to date at the time of publication, but individual items may not remain so.

GENERAL

The Institute of Patentees and Inventors (IPI)
505a Triumph House
Regent Street
London W1
Telephone: 071-242 7812

Anyone can join, whether they have an invention or not. Fees are £55 for the first year, £43 after that. Discount rate for students. See page 210.

The Design Council
Innovation Service
28 Haymarket
London SW1Y 4SU
Telephone: 071-839 8000
Fax: 071-925 2130

As noted in Chapter Nineteen, the Council's Innovation Service is to be handed over to other organizations (as yet unannounced). Should you discover, on getting in touch with the Council to ask for help, that it no longer runs Innovation Services, I suggest that you try the Department of Trade and Industry's Innovation Enquiry or Business Link departments (see page 213).

Meanwhile, I report the position which exists as this book goes to press. The Council offers free of charge a superb information pack. No inventor can afford to be without it. Its *Contacts Directory* is the most comprehensive available (except this one), listing every form of regional assistance for innovators, plus commercial help, and libraries and information centres. It gives information on prototypes, product development, marketing and promotion, design, licensing and sources of finance and venture capital.

NoticeBoard is another free service. Its idea is to help innovators get their projects up and running by matching them up with companies on its database (see page 212).

The Product Development Advisory Service (PDAS) gives advice on property rights, design protection, sources of finance, licensing and setting up in business.

The Enterprise Initiative Design Consultancy (EIDC) is a scheme operated on behalf of the DTI. It lets companies with design projects use a specialist consultancy for up to fifteen days, with at least half of the costs paid to the DTI. The Council says that projects can involve anything: electronics and software design, engineering design and analysis, and product packaging. Note: the fate of EIDC is definitely known. It is being moved to the DTI's Business Link network, which is still in the process of creation.

The Department of Trade and Industry (DTI)
No address applicable
See below for Freephone numbers

Freephone telephone numbers are the initial point of contact between the public and the central body of the DTI. They can tell what the Department is doing, as well as helping to point inquirers in the right direction. The DTI is very decentralized and works through its regional offices. These are not always easy to track down (though local reference libraries should know where they are, as should the nearest Chamber of Commerce). But the quick way to get an answer is to ring the Freephone numbers.

The DTI is a government department. Schemes for innovators and small businessmen may change or be withdrawn at any time. It runs Enterprise Initiative, Innovation Enquiry and Business Link.

Enterprise Initiative is a package which offers many delights. These include five to fifteen days of consultancy in key business areas such as planning, design, financial management, manufacturing and services systems, marketing and quality. The DTI pays one third of the costs (half for firms in special areas) and almost all companies with fewer than 500 employees are eligible. There is help with solving technical problems, grant support to develop innovative ideas and special assistance for companies in Assisted Areas and Urban Programme areas. For further information, call Freephone 0800 500200.

For information on Innovation Enquiry, ring Freephone 0800 442001. The DTI runs the SMART and SPUR schemes which offer grants to innovators and small firms (see page 213–14).

Business Link is a new government strategy to deliver a wide range of services to businesses through a series of local offices. There are a couple of dozen such offices open and eventually there will be 200. Innovation Enquiry will probably be subsumed into Business Link, as may some of the functions of the Design Council.

The DTI is pursuing its policy of decentralization and you should find your nearest Business Link office through the phone book, the local library or the council. It does not seem likely that the offices will have much to offer innovators as such (unless they distribute the Design Council's *Contacts Directory*) but they may be of value to the inventor who is already in business.

PATENTS

The Patent Office is an executive agency of the Department of Trade and Industry.

Marketing, Publicity and Information Service
The Patent Office
Room 1L02
Cardiff Road

Newport
Gwent NP9 1RH
Telephone: 0633 813535
Fax: 0633 813600

Everything you need to know about patents and intellectual property, how to prepare your patent application and process it through the system, trade and service marks, an introduction to copyright, and the search services the Patent Office can offer. The information is all free.

The Patent Office
Cardiff Road
Newport
Gwent NP9 1RH
Telephone: 0633 814611

The heart of the matter. This is where you prepare and file your patent applications, and make payments for them. The address is the same as the Patent Office Marketing, Publicity and Information Service, but the telephone number differs.

Patents can also be filed in London at:

The Patent Office
25 Southampton Buildings
Chancery Lane
London WC2A 1AY
Telephone: 071-438 4700

Southampton Buildings, like Newport, accepts patent applications of all kind, not only for the UK but European and Patent Co-operation Treaty countries (see Chapter Seven). It also handles trade marks and Registered Designs.

Search and Advisory Service
The Patent Office
Hazlitt House
45 Southampton Buildings
Chancery Lane
London WC2A 1AR
Telephone: 071-438 4747/8
Fax: 071-438 4750

Generally speaking, the Search and Advisory Service is a commercial facility most useful to established companies planning to spend sums on R and D projects. It has the assistance of over 200 patent examiners, each of them a specialist in a particular area of technology and innovation. They can check through existing patents, to find if a new product or process is likely to win patent protection. 'Family' searches detect similar patents published by other countries and international authorities.

Examiners identify where making or using a product might breach a patent still in force. They can be employed to keep an eye on the patenting activities of competitors.

One of the Service's specialities is the field of chemicals. It has a search and display system called CAS ONLINE. This examines a computerized database of more than ten million substances.

You don't have to file for patent protection to employ the Service. It will accept commissions sent in by letter, phone, telex or fax. But it does suggest that a personal visit is useful. Charges and estimates are available on request.

This is big stuff and usually not appropriate for the private inventor, unless he or she has a lot of money. Don't worry. You can carry out your own free search in the Science Reference and Information Service library (see below).

Science Reference and Information Service (SRIS)
The British Library
25 Southampton Buildings
Chancery Lane
London WC2A 1AY
Telephone: 071-323 7919/7920 for British and EPO Patent inquiries.
Opening hours: Monday to Friday 0930 to 2100; Saturday 1000 to 1300.

The national patent library. UK and European patents are in the Holborn Reading Room. Foreign patents, and trade marks, are in Chancery House Reading Room. The SRIS moves to the new British Library building in Euston Road when that is opened.

The SRIS library is vitally important for inventors, as it has the most complete reference collection in Europe of literature on patents, trade marks and designs, and holds over thirty-three million patent specifications from countries round the world.

Personal visits are vital to use the manual search system, which is free. (See Chapter Five.)

The library also offers fee-based services:

Patents Online
Telephone: 071-323 7903
Fax: 071-323 7480

Research by library staff. They use the major commercially available patent databases.

Patent Express
Telephone: 071-323 7926–9
Fax: 071-323 7930

Supplies copies of patent documents within forty-eight hours, or three hours 'Rush' service.

Transcript
Telephone: 071-323 7929
Fax: 071-323 7930

Competitively priced English-language translations of any patent specification in the library.

PROVINCIAL PATENT LIBRARIES

There are eight of these, providing a useful service for the large percentage of the population which finds it difficult and expensive to visit London. They are, naturally, not as comprehensive as the SRIS library in London but you can hope to find UK, European and some foreign patents, especially American, together with patent journals and indices.

Manual searching is free, but paid search services are also available. The advice of library staff can be invaluable. Ring to find out what they have in stock before you invest time and trouble in a personal visit.

Belfast
Central Library, Royal Avenue,
Belfast BT1 1EA
Telephone: 0232 243233
Birmingham
Patent Department, 5th Floor, Central Library,
Chamberlain Square, Birmingham B3 3HQ
Telephone: 021-235 4537/8
Glasgow
Mitchell Library, North Street,
Glasgow G3 7DN
Telephone: 041-221 7130 ext 279

Leeds
Patents Information Unit, Library HQ,
32 York Street, Leeds LS9 8TD
Telephone: 0532 488747
Liverpool
Patent Library, Science and Technology Library,
William Brown Street, Liverpool L3 8EW
Telephone: 051-225 5442
Manchester
Patent Library, Central Library, St Peter's Square,
Manchester M2 5PD
Telephone: 061-236 9422
Newcastle-upon-Tyne
Patents Advice Centre, Floor A, Central Library,
Princess Square, Newcastle-upon-Tyne NE99 1DX
Telephone: 091-232 4601
Sheffield
Central Library, Surrey Street, Sheffield S1 1XZ
Telephone: 0742 734742

The Chartered Institute of Patent Agents
Staple Inn Buildings
High Holborn
London WC1V 7PZ
Telephone: 071-405 9450

The Institute offers three clear and useful booklets, free of charge: *Inventions, Patents and Patent Agents*; *Industrial Designs, Copyright and Patent Agents*; *Trade Marks, Registered Trade Marks and Patent Agents*. It publishes the *Directory of Patent Agents*, organized county by county, except for Scotland and Wales, which have a single section each. Many of the agents listed are also qualified as European Patent Attorneys and can represent patent applicants before the European Patent Office.

Finally, there is a four-page leaflet, *Exploitation of Inventions*, which gives the questions you should ask yourself in the early stages of deciding how to protect your invention. It suggests some typical costs e.g. upwards of £2000 by the time a British patent has been granted, using the services of an agent.

Patents abroad
European Patent Office
Erhardstrasse 27
D-80298 München

Germany
Telephone: 010 49 89 23990
Fax: 010 49 89 2399 4465

European Patent Office
Patentlaan 2
Postbus 5818
NL 2280 HV Rijswijk
Netherlands
Telephone: 010 31 70 3400 2040
Fax: 010 31 70 340-3016

European patents can protect an invention in seventeen countries, from Sweden to Italy, from Portugal to Austria – in fact, the whole of western Europe, except Norway and Finland but including Greece.

Applications may be filed with the EPO in either English, French or German, at the Munich office itself, at The Hague or at the UK Patent Office. Europatents are valid for twenty years from the date on which the application is filed.

The EPO issues a *Guide for Applicants* which covers the whole field of patents from the concept of Patentability, through Drawing Up and Filing an Application to an analysis of the Procedures for granting patents.

Charges and fees are published regularly in the *Official Journal of the European Patent Office*.

See Chapter Seven for more about European Patents.

US Information Service
24 Grosvenor Square
London W1A 1AE
Telephone: 071-499 9000

Staff are helpful in the nicest American manner and will even answer straightforward inquiries (such as addresses of American companies) over the phone, but to research properly you have to visit in person. The library is for reference only.

There are official publications on commercial subjects, US business magazines and trade journals, directories and annual reports on leading American companies. If you are thinking of patenting in the USA or exporting to them, inquiries start here.

The US Embassy itself has a section for Economic and Commercial affairs: telephone 071-408 8020. And the Embassy, like nearly all those belonging to industrialized nations, has a Commercial Attaché on the strength.

Diplomacy is as much about trade as it is about foreign policy and these attachés work hard to promote good commercial relations between countries. They can be extremely helpful in putting you in touch with the right people or organizations in their own countries.

World Intellectual Property Organization (WIPO)
34 Chemin des Colombettes
1211, Geneva 20
Switzerland
Telephone: 010 22 99 91 11

WIPO is a United Nations agency. It looks after the protection of intellectual property worldwide through agreements such as the Paris Convention or the Patent Co-operation Treaty (see Chapter Seven). You aren't likely to get involved.

COPYRIGHT

Intellectual Property Policy Directorate
Patent Office
Hazlitt House
45 Southampton Buildings
Chancery Lane
London WC2A 1AR
Telephone: 071-438 4778 (copyright inquiries)

How do you get permission to use copyright material which belongs to someone else? Is your own copyright protected in the countries where your product is on sale? Where does copyright exist with computer software? Don't ring the Directorate until you have read the Patent Office booklet *Basic Facts – Copyright*. See Chapter Nine.

DESIGN

Designs Registry
The Patent Office
Cardiff Road
Newport
Gwent NP9 1RH
Telephone: 0633-812515

Are you able to register the design of your idea? How does a design differ from a patent? What does it cost? What is design right and what protection does it give? Before you call them, read the Patent Office booklet *Basic Facts – Designs*. See Chapter Nine.

TRADE MARKS

Trade Marks Inquiries
The Patent Office
Cardiff Road
Newport
Gwent NP9 1RH
Telephone: 0633-814706

What is the difference between a trade and a service mark? How can you register? What do they cost and what advantage do they give you?

Read the Patent Office booklet *Basic Facts – Registered Trade and Service Marks* before you ring with inquiries. See Chapter Nine.

Trade Marks Search and Advisory Service
The Patent Office
Cardiff Road
Newport
Gwent NP9 1RH
Telephone: 0633-814700

A commercial and computerized service which uses the experience of seventy trained examiners. It is available independently of applications to register a trade or service mark. Searchers investigate the proposed mark to discover if it is likely to be confused with any existing one. Hearing Officers advise on ways to change proposed designs so that conflict does not arise.

The Service will even search for designs containing heraldic devices and (if you have the agreement of the Kings of Arms) advise on whether they can be registered as marks!

Search requests are accepted by letter, telephone, fax or personal call. Costs on request.

Institute of Trade Mark Agents
Canterbury House
2–6 Sydenham Road
Croydon
Surrey CR0 9XE
Telephone: 081-686 2052
Fax: 081-680 5723

Founded in 1934, the Institute is responsible for administering the Register of Trade Mark Agents created by the 1988 Copyright, Designs and Patents Act. If you should need an agent, the Institute has a list of several hundred members, both in the UK and overseas.

It also offers an explanatory pamphlet, *Trade Marks – An Introduction*, and a leaflet which sets out the scale of professional fees. For example, filing a UK application plus paying the registration fee costs £385.

STANDARDS

British Standards Institution (BSI)
(Enquiries Section)
Linford Wood
Milton Keynes MK14 6LE
Telephone: 0908 220908
Fax: 0908 220671

It always surprises me that, except where there is specific legislation – usually for reasons of safety – the observation of standards is voluntary. Manufacturers don't have to use them. Compulsory or not, widely accepted standards are your best guarantee to customers that what they are buying comes up to scratch. BSI was set up in 1901 to provide such standards. Eighty per cent of BSI Standards income derives from subscriptions and sales; the rest comes from the government.

BSI will explain how your product can gain the right to carry its Kitemark or Safety Mark (which have considerable advertising and marketing value). It will run tests to assess your product's compliance with various British or international standards. It charges for these tests.

These days there is even a standard for assuring quality. BS 5750 covers every factor affecting quality in every part of a company. More than 7000 companies are registered.

The whole thing can seem a little intimidating to a lone entrepreneur. BSI is conscious of this and has a new programme to assist and inform small businesses. It hopes to clear away some of the mystique surrounding standards.

LICENSING

Institute of International Licensing Practitioners Ltd
28 Buckingham Palace Road
London SW1W 0RJ
Telephone: 071-630 0245
Fax: 071-630 0235

The Institute's telephone number is a first contact. Explain your problem and you should be put in touch with someone who can handle your query. Half of the Institute's members are in the UK, the rest around the world. It keeps a register of fellows and associate members, with their qualifications and areas of special knowledge. This will be sent on request. See Chapter Eleven.

MARKETING

The Chartered Institute of Marketing
Moor Hall
Cookham
Berkshire SL6 9QH
Telephone: 0628 524922
Fax: 0628 531382

Marketing consultants are best used in very competitive situations. If you are aiming for a mass market, they can be of help.

The Institute operates a Managed Business Consultancy Service (MBCS). It is available to all businesses of any size, anywhere in the world. Contact: Director of Consultancy Services; telephone: 0628 852150.

Each client is allocated a Project Manager who matches the client's marketing needs to a relevant consultant to create a working partnership. After the terms for this are agreed, the Project Manager follows through, checking quality control and making sure that the client receives maximum benefit from the consultancy. When the final report is presented, it is discussed by all three parties and only when the client is completely satisfied with the report is the project signed off.

Another service offered by the Institute is INFOMARK, telephone 0628 852190. This is a unique marketing library and information service, available to non-members as well as members and students, which has 4000 texts, journals, directories, statistical information and a customized key search database. It covers products and services; retail, wholesale and distribution; company information; and every aspect of marketing.

INNOVATION CENTRES

Have you an idea which you would like to develop, but you don't know how? Do you need to build a proper working model? Would you like hands-on help with questions of technical and commercial evaluation? How about patent and trade mark searches, or assistance with licensing agreements?

There are Innovation Centres around the UK which can help you with some, if not all, of these questions. Some are funded locally by Chambers of Commerce, some are helped by government money, some are funded by the European Business and Innovation Network (EBIN), through the European Union. Below they are in one list with members of EBIN starred.

Innovation Centres offer no miracle route to the success of your idea. Profit is *not* guaranteed. But they are good value for the private innovator. Try the following:

*BARNSLEY BUSINESS AND INNOVATION CENTRE
Barnsley, South Yorkshire S75 1JL
Telephone: 0226 249590
Fax: 0226 249625

*BIRMINGHAM TECHNOLOGY LIMITED
Birmingham B7 4BJ
Telephone: 021-359 0981
Fax: 021-359 0433

BUSINESS DEVELOPMENT CENTRE
Telford, Shropshire TF3 3BA
Telephone: 0952 290769

CORNWALL INNOVATION CENTRE
Camborne TR14 0AB
Telephone: 0209 612670
Fax: 0209 612671

DESIGN WORKS (GATESHEAD)
Gateshead NE10 0JP
Telephone: 091-495 0066
Fax: 091-495 3207

DONBAC
Doncaster DN1 3NA
Telephone: 0302 340320
Fax: 0302 344740

ENTERPRISE PLYMOUTH
Plymouth PL3 4BB
Telephone: 0752 569211
Fax: 0752 6095250

GLASGOW OPPORTUNITIES
Glasgow G2 1EQ
Telephone: 041-221 0955

*GREATER MANCHESTER BUSINESS INNOVATION CENTRE
Greater Manchester M34 3QS
Telephone: 061-337 8648
Fax: 061-337 8651

HEREFORD AND WORCESTER TEC
Worcester WR1 1UW
Telephone: 0905 723200
Fax: 0905 613338

*INNOVATION CENTRE NORTHERN IRELAND
Londonderry BT48 0NA
Telephone: 0504 264242
Fax: 0504 269025

*INNOVATION WALES
Cardiff CF2 4AY
Telephone: 0222 372311
Fax: 0222 373436

*LANCASHIRE BUSINESS AND INNOVATION CENTRE
Blackburn, Lancs BB1 3BL
Telephone: 0254 692692
Fax: 0254 692290

LONDON ENTERPRISE AGENCY
London EC1A 2BS
Telephone: 071-236 3000

MANCHESTER BUSINESS VENTURE
Manchester M60
Telephone: 061-236 0153
Fax: 061-236 4160

MEDWAY ENTERPRISE CENTRE
Gillingham, Kent ME8 0RW
Telephone: 0634 366565

MERSEYSIDE INNOVATION CENTRE
Liverpool L3 5TF
Telephone: 051-708 0123

MILTON KEYNES BUSINESS VENTURE TRUST
Milton Keynes MK9 2AE
Telephone: 0908 660044
Fax: 0908 233087

*NEWTECH INNOVATION CENTRE
Deeside, Clwyd CH5 2NT
Telephone: 0244-289881
Fax: 0244-280002

NORTH EAST INNOVATION CENTRE
Gateshead NE8 3AH
Telephone: 091-490 1222

*NOTTINGHAM CENTRE FOR BUSINESS AND INNOVATION
Nottingham NG7 2QP
Telephone: 0602 436643
Fax: 0602 220718

OGWR PARTNERSHIP TRUST
Bridgend, Wales CF32 9BS
Telephone: 0656 724414
Fax: 0656 721163

SANDWELL ENTERPRISE
West Bromwich B70 8ET
Telephone: 021-500 5412

*SCOTTISH INNOVATION
Glasgow G40 1DA
Telephone: 041-544 5995
Fax: 041-556 6320

SOMERSET INNOVATION CENTRE
Taunton TA1 4AS
Telephone: 0823 276905

SOUTH LONDON INNOVATION CENTRE
London SW2 1HJ
Telephone: 081-671 4055

STRATHCLYDE REGIONAL COUNCIL
Glasgow G3 8XA
Telephone: 041-227 2243
See Chapter Six.

Innovation Services

Imagineering
Bury Road
Ramsey, Huntingdon
Cambridgeshire PE17 1NE
Telephone: 0487 813336

Supplies practical advice about innovation to small and medium-size firms, and starter companies. Imagineering will develop new products, carrying them through from their conception to placing them with manufacturers and distributors.

Imagineering has specialist experience in the fields of general engineering, agriculture, road transport and computer animation. The service is fee-based; Imagineering takes no financial cut from the success of any invention which it midwifes. It works to a Code of Practice. See Chapter Four.

Competitions and Awards

Livewire
Hawthorn House
Forth Banks
Newcastle-upon-Tyne NE1 3SG
Telephone: 091-261 5584
Fax: 091-261 1910

Livewire aims to assist young people interested in starting their own businesses. Its main backer is Shell UK. The scheme is organized regionally, with over eighty local co-ordinators.

It says: 'A young inquirer can feel assured that a Livewire adviser will be prepared to listen to them and not write them off because of age, appearance or a "crazy" idea. Livewire aims to encourage any young person to think about the option of self-employment, and then help with commonsense and business acumen.'

Livewire has three programmes:

Enquiry and Link Up Service
This offers UK youngsters from sixteen to twenty-five who want to start their own business, free local advice and support. The booklet *Unlock Your Potential* shows how to take the first steps and, if you ask for one, a suitable local adviser will be appointed. There is a flexible Business Plan Guideline available to every inquirer.

Business Start Up Awards
Young business people, again in the sixteen to twenty-five age group, compete at county and regional levels on their way to the Livewire UK final. They have to submit a business plan to their local co-ordinator. There are awards worth £175,000, plus support 'in kind'. A major cash prize is given to the most promising young business at the UK awards ceremony.

Business Growth Challenge
This is for people or companies, employing up to four staff, who have been trading for over eighteen months and are looking to the next stage in developing their business. Successful applicants are offered places on a residential weekend. This combines a series of individual and team-based challenges, designed to develop management skills and personal qualities. Afterwards, participants are encouraged to come up with a plan for the future of their businesses and they can be given further advice or training.

The Prince of Wales Award for Innovation
c/o Business in the Community
8–9 Stratton Street
London W1X 5FD
Telephone: 071-629 1600
Fax: 071-629 1834

Launched in 1981 to find new technological ideas and encourage them towards commercial production, this award became associated with the Prince of Wales in 1989. It is perhaps the award with most prestige, because of the Royal connection (see page 214).

Rotary Young Inventor of the Year
Rotary International in Great Britain and Ireland
Kinwarton Road
Alcester
Warwickshire B49 6BP
Telephone: 0789 765411

The contest offers a tremendous first prize of £10,000. The winning invention must show quality in its presentation and potential for commercial development but, above all, inventiveness. By this, the organizers mean that they are looking for originality, simplicity and effectiveness of design, and safety in use.

Young people between the ages of eleven and eighteen take

part; group entries are not accepted. Rotary expects that most contestants will come direct from schools and asks youngsters with a new idea to submit it through their CDT teachers, or to the local Rotary club. If this isn't possible, call the number given above for further information.

The contest is organized on a regional basis, first at the local Rotary Club level, then at district level (Rotary in Great Britain and Ireland is divided into twenty-nine Districts) and finally with a national Grand Final. See page 217.

Shell Technology Enterprise Programme (STEP)
11 St Bride Street
London EC4A 4AT
Telephone: 071-936 3556

STEP is designed to help independent companies with fewer than 200 employees which have projects of a commercial or technical nature that can be completed by penultimate-year undergraduates in eight weeks i.e. during their summer vacation. By reflection, it offers young people a chance to show their mettle in the real world of innovative business. They are paid a weekly allowance of £100 a week while they do so. See page 215.

Young Engineers for Britain
The Engineering Council
10 Maltravers Street
London WC2R 3ER
Telephone: 071-240 7891

An annual contest for young people either in full-time education or in industry. Essentially a regional competition, held in a dozen centres in Britain. Prizes up to £20,000 in total. See page 216.

Young Engineers
Dept of Educational Studies
University of Surrey
Guildford
Surrey GU2 5XH
Telephone: 0483 259349

Students in schools' Young Engineers clubs, helped by engineers from industry, devise solutions to real-life problems. The scheme has an annual National Award ceremony with prizes totalling over £60,000. See page 217.

Index